The Devil's Reign

A Documented True Story That Proves The Forces of Good and Evil Do Exist

By
George Newberry

Bloomington, IN Milton Keynes, UK

authorHOUSE®

AuthorHouse™
1663 Liberty Drive, Suite 200
Bloomington, IN 47403
www.authorhouse.com
Phone: 1-800-839-8640

AuthorHouse™ UK Ltd.
500 Avebury Boulevard
Central Milton Keynes, MK9 2BE
www.authorhouse.co.uk
Phone: 08001974150

First published by AuthorHouse 4/4/2007

ISBN:1-4208-5398-8 (sc)
ISBN: 1-4208-5397-X (dj)

Printed in the United States of America
Bloomington, Indiana

This book is printed on acid-free paper.

Contact the author at GeoNewberry@aol.com

**This book is dedicated to the memory
of my mother and stepfather.**

**Thomas and Edith Murphy
Taylorville, Illinois**

With you no longer here to share life with me,
the sweetness of success can never be as fulfilling.
My life will never be the same without you.

PREFACE

The Devil's Reign is a true story. The statements made by the author in this book have been validated with documentation and verified by this publisher's legal department.

You will read about unexplainable, mysterious, supernatural, mind-boggling events. You will find them hard to believe, but everything within these pages is validated by witnesses or legal documentation. These are the experiences and events that have molded me into the person I am.

This is a story about a near-death experience, a dream, and events that occurred as a direct result of my efforts to pursue my mission. From my experiences I have learned that the forces of good and evil do exist.

Is the Devil's time short? Is Satan in control of the world? Is he spreading his evil everywhere trying to win every soul he can get before his time is up? Are we being warned, no matter what religion we may be, to get our lives in order with God before it's too late? Based on my experiences, I believe so.

I am not a religious fanatic. I am no better than any ordinary human being. I am far from being perfect and I am a sinner like everyone else. I have wondered many times why I, of all people, was chosen to write this story. But for whatever reason, some things have happened to me that, to my knowledge, have not happened to anyone else. Spiritual things. Supernatural things. Things that are hard to believe. Things that have placed me in direct conflict with an evil force: The Devil, Satan, Lucifer, The Force of Evil or whatever you may call him. Satan doesn't want this story told. He has done his best to destroy both my story and me. My faith in God and my mission gives me the strength to withstand adversity and keeps me from giving up.

What is my destiny? I don't know all of it. I have heard no divine voices. I have not had a conversation with God. But I have had many experiences that have given me a story that I know is my destiny to tell. It is a story that may help stop evil and change people's lives for the better.

The biggest trick of Satan is convincing people he doesn't exist. He also wants people who do believe he exists to be afraid of him. He definitely doesn't want people to believe there is a God. And, if they do believe there is a God, he wants them to blame God for all their problems.

From the time I began to fulfill this "destiny," I have been under attack by an evil force. I have dealt with tremendous heartaches, gut wrenching anxieties, and huge financial losses. It hasn't been easy. But I have survived. People pretending to be friends, who may have been unaware they were being used by Satan, caused me the most pain.

I know the battle isn't over; I must keep fighting, regardless of what happens to me. I can't give up. I can't let evil win. I must keep my faith in God, faith in what I know and believe, and faith in what I am doing. Only the things that God wants to happen will happen.

All people are born with a destiny. If I am truly fortunate enough to have been given a special destiny, I pray that I successfully fulfill it. Regardless of what is taking place in my life or in the world in which we live, in the end God will triumph. There will be peace on Earth.

I can only tell the story of what has happened to me. I can only interpret what it means to me. Though witnesses and documentation prove this story is true, there will be people who will choose to not believe the witnesses, the documentation, or me. That is ok. We all have the right to choose what we do, or do not, believe. But regardless of positive or negative feedback, my mission can only be fulfilled if I tell the truth. I hope reading this story changes your life for the better.

The time will come when we will all know who is right, who is wrong, what is true, what is false, what is fact, and what is fiction. Regardless of how our religious or spiritual beliefs may differ, *evil is our common enemy*. All religions must unite forces and fight evil together, or we will all be destroyed. We are in the midst of a difficult battle that will only cease when people, or God, bring an end to "**The Devil's Reign**."

CHAPTER 1

The war between the forces of good and evil has begun. The battle lines have been drawn. Satan's armies are on the march. A wave of evil is inundating the world. The forces of evil are powerful and relentless. They won't stop their assault until they achieve complete victory over mankind, or until they are defeated. If we sit in our homes or churches and feel safe because we pray for God to stop evil, we are fools. It is not God's place to stop evil. People create evil, people allow evil to flourish, and people must stop evil. If we are destroyed, it will be by our own hand. We are living during The Devil's Reign but we possess the power to bring it to an end.

If I had known I was going to become entangled in a battle between the forces of good and evil, I would have pursued my mission with fear and trepidation. I didn't realize my efforts to warn mankind to fight ungodliness would cause me to be pursued by the Ultimate Force of Evil.

I have been the recipient of many unexplainable occurrences. Some transpired recently, some happened not long ago, and some took place when I was a child. I had a near-death experience at the age of fourteen.

When my mission first began, I was reluctant to tell people about my experiences. I feared that people wouldn't believe me. If they scoffed or laughed I was intimidated. That is no longer the case. My experiences have made me a wiser and stronger person. I must tell the truth regardless of what people think or say. If they choose to scoff and laugh, that is ok. Those who choose to read this story with a receptive mind will gain much. They will know the forces of good and evil exist.

I have furnished declarations from witnesses and legal documentation to the legal department of my publisher to back up the statements I have made in this book. I welcome an investigation into anything I have written. My friends who have witnessed these events, and I, are willing to submit to polygraph tests.

My story officially began November 21, 1940. That was the day I was born in the small town of Wyatt, Missouri.

Wyatt is located a short distance from the Mississippi River at the junction of Missouri, Kentucky, and Illinois. In the 1940's, Wyatt was a typical small southern town. Only a few hundred people lived there and the majority of them were black. The main business was the cotton gin and most of the people picked cotton for a living. Most of the homes had no electricity, gas, or indoor plumbing. Homes without those luxuries had an outhouse. If you had a double wide with two holes you were styling.

I was the fourth child of five born to my mother Edith Lawrence Newberry. The first and third, a sister and a brother, both died of pneumonia. The second child, my sister Mary, survived. I arrived four years later. Two years after that we were blessed with the birth of my younger sister Jean.

Shortly after Jean was born, my father enlisted in the army to fight in World War II. My mother, like many others, was left struggling to raise her kids. The War was going full blast and times were hard. Not unlike most of the residents of Wyatt, we were very poor. We lived in a shotgun house, three rooms all in a row. Our house, like most others, had no electricity, indoor plumbing or sewer. We made many miserable trips through snow and rain to the outhouse. Life was simple. Everything had to be done the hard way. Our humble lives were a far cry from living in luxury. But that didn't matter to us. We loved one another. We had a wonderful family and we were happy.

Mom had to take my sisters and me with her almost everywhere she went. We loved going to the grocery store with her. Fred and Sylvia Burgstaller's Grocery Store was a dilapidated old building. It was covered with deteriorating fake brick siding and wood that was devoid of paint. Like an old general store, groceries were on shelves behind the counter. Barrels of grain and bags of flour covered the floor. The store was always filled with black people who stopped in regularly to see Fred. Many of them sat on the barrels and flour bags and made the store their daily 1940's "coffeehouse." They were kind jovial people and they were liked by everyone. Fred always teased them and they always laughed and teased him back. Fred liked them and they liked Fred.

Fred and Sylvia were good people. They lived only a step higher up the ladder than most of the townspeople, but they owned a store and to everyone else they were rich.

During World War II, lots of items were in short supply and hard to get. The government rationed many of them. If you didn't have a ration stamp for that item, you didn't get it. My mother was young, blonde, and probably the prettiest lady in the town. Fred always flirted with her and Sylvia was one of her best friends. They knew my mother needed help and they were good to her. Thanks to them, we always got everything we needed. And, my sisters and I always went home with free lollipops.

At that time the country was filled with racism. But, my mother taught us to never be racist or judgmental of others. I especially remember one time when we were walking home from Fred's store. A black man, who was approaching us on the sidewalk, stepped into the street to let us pass by.

"Excuse me, Ma'am," he said to my mother.

"Hey," my mother replied. "You get back on that sidewalk. You don't have to step out of the way for me. You have as much right to be there as I do."

The man was hesitant to step back onto the sidewalk but my mother insisted that he did. I can remember his smile and the gleam of appreciation his face showed for my mother.

To wash clothes, Mom heated water in a big iron kettle over a fire in the back yard. That kettle, plus a wash board, was her washing machine. It's a miracle she didn't wear her knuckles off on the wash board. I remember standing in the kitchen making airplanes from clothespins while I watched her heat a heavy old iron on a wood-burning kitchen stove. Then, she pressed clothes with one hand and wiped sweat off her brow with the other.

My mother was a true survivor. Now, most people couldn't cope with the problems and the struggles a lot of people dealt with on a daily basis back then.

Like most of the people who lived in Wyatt, Mom picked cotton to make a living. Childcare facilities didn't exist and she was too poor to pay a babysitter, so she always took my sisters and me to the field with her.

Mom picked the cotton and put it in a long shoulder bag she dragged on the ground behind her. When the bag was full of cotton it was so heavy all of us struggled to drag it to the side of the field and dump it into a wagon. Then my sisters and I climbed into the wagon and jumped up and down to pack the cotton and make room for more. I don't remember how many bags of cotton Mom picked per day, but she was a hard worker. When she went home for the night she was exhausted and her hands bled from being pricked and scratched by stickers on the cotton bolls. She didn't care about that, she was only concerned about her kids. Money can't buy memories like those, and they are a part of the reason I love my mother so much. Her family was her life.

I can't remember much about my father, William Sanford Newberry. His family and friends called him Willie. He fought in some of the worst battles of World War II in North Africa and Italy, but when the war was over, he came home without a scratch.

I vividly remember the day my father came home. My sister, Jean, and I were in the front yard playing with a little red wooden chair, when we saw a man dressed in a military uniform walking toward the house. I had been too young to remember seeing my father before he enlisted in the army, but I instantly knew who the man was. We were all thrilled he was home. Mom was so happy she cried.

We soon became a normal family living a normal family life. My sisters and I had a father and my mother again had a husband. She was able to have life a little easier. But that only lasted for a few short weeks. One day, my father went to a local bar with some of his friends. He hadn't been gone long when Mom became very distraught. She kept pacing the floor, back and forth, repeating over and over, "I know something is wrong, I know something is wrong."

She was a nervous wreck. She finally asked a neighbor lady to stay with us, and she went looking for my father. Within an hour she returned home. She was unable to find him. However, she had seen one of his friends. That friend said he had just seen my father drinking with some of his friends and he was fine. That didn't satisfy my mother and she remained very uneasy.

I was playing in the front yard when a dark green car pulled up in front of the house. My mother came running outside, yelling, "Where is he? I know something is wrong. Where is he?"

"There's just been a little accident," the driver responded.

My mother was very distraught. She got into the car and the car sped away. Juanita, the neighbor lady my mother had watching us, heard the scene of the accident was only a few blocks away. Instead of staying at home with my sisters and me, Juanita wanted to go see the accident. I still don't understand why she took us with her.

I remember feeling a bit apprehensive as we walked to the site. I was only four years old, but I knew something was terribly wrong. I had an idea of what had happened and I knew we weren't going to see anything pleasant. When we arrived at the site we pushed our way through the crowd and came face to face with carnage no child should ever see. Blood was everywhere. A mangled car was strewn on the roadside and a hubcap was embedded in the trunk of a tree.

The driver of the car had only minor injuries. He was driving drunk, lost control of the car, and hit the tree. My father was lying on the ground. His head was propped up on a car seat that had been thrown from the car. His face was almost torn off and he was covered with blood. He had been ejected from the car, head first, into the tree. He was moving, but he was barely alive.

My mother was crying hysterically and screaming for God to help him. She was trying to hold him in her arms and her dress was covered with his blood. Some of the people were trying to console her. An ambulance arrived and backed up close to my father. The drivers placed him on a gurney and rolled him inside the ambulance. My mother climbed into the ambulance beside him and continued to weep. Someone closed the door and the ambulance sped away.

That was the last time I saw my father alive. He had fought in some of the worst battles in North Africa and Italy, and he had come home unscathed. Now, only a few weeks later, he was dead. The only memory of any interaction I had with him was when he gave me a dime to go to Burgstaller's store and buy a toothbrush.

My mother had a hard time coping with his death. I look back now and wonder how she kept her sanity. Within just a few years she had lost a daughter, a son and her husband.

I was too young to realize the seriousness of what had happened but I vividly remember the flag draped coffin at the funeral home and the funeral services at the church. At the cemetery we sat under a green tent that covered the gravesite. It rained during the entire graveside service. After the minister said his prayers, I watched some men remove the flag from my father's casket. They folded it and handed it to my mother. I intently watched the coffin as it was lowered into the ground, and I listened to my mother cry.

Sometimes we have a hard time understanding why God lets things happen the way they do. Life is filled with pain and sorrow, good and bad, happiness and tears. Life isn't always easy, but there is a reason for every experience.

CHAPTER 2

World War II was over and troops were returning home. One soldier returned to the house across the street. His name was Thomas Agustus Murphy. Everyone called him Tom or Gus. His wife had passed away a few years earlier. Gus, the name I always called him, was left with a son who was two months older than I was. Mom and Gus started dating and I soon had a new father and a new brother named Edward. Shortly after their wedding we moved from Missouri to Taylorville, Illinois.

We were ecstatic when we moved into the house in Taylorville. We had electricity and plumbing. An addition was soon built onto the house and we even had an indoor bathroom. I'll never forget the first Christmas we had in that house. We had one big blue light bulb on our tree, and we were so excited we stayed up all night just to look at it.

We were only in Taylorville about a year when we had a new addition to the family, my brother, Tommy. Now there were seven of us living in that house, with her kids, his kid and their kid. We didn't have everything we wanted, but our needs were satisfied and we were happy.

Soon, Gus got a good job and things got better and better. We got new furniture, a new car and another addition was built onto the house. We felt like we were rich. I wouldn't trade the way I was raised for anything. I am so lucky to have the memories of those times. We always have been and always will be a close family. Our parents were strict disciplinarians, but we respected them and knew we deserved every punishment we received. Well, almost. All in all, our experiences and the way we are raised are things that build our character and our moral and spiritual values. That is why I wouldn't change a thing that has happened in my life, and why I think it is necessary to include these things in this story.

CHAPTER 3

Life was about the same for the next few years, that is, until the summer of 1955. I was fourteen years old and ready to go into the ninth grade. In August, during summer break, I spent a fun-filled day at the County Fair with friends. What a good time I had. I think I rode every thrill ride there. I was spun around, shook, flipped upside down and jolted all afternoon. It was loads of fun, but I think it was the catalyst that started the nightmare that was to follow.

In the early evening my family was having a backyard barbecue. I filled my plate and went into the house. "Topper," a popular comedy series, was on television and I couldn't miss it. The show was funny and I began to laugh. Suddenly, I felt a strange tickle in my throat. It was unlike anything I had felt before. I started coughing and I felt something warm in my mouth. I ran to the bathroom to spit out whatever was there. To my dismay, it was blood.

I don't know why, but I wasn't frightened. We were leaving on vacation the next day, and I didn't want to spoil the trip. So, I decided not to tell anyone what was happening.

When I went to bed, every time I would lie down I coughed up more blood. So, I sat up all night and spit the blood into Kleenex. By the next morning the wastebasket was full.

I didn't realize I had a serious problem, and I still didn't tell anyone what was happening. We were going to Tennessee to visit my mother's sister, and I wanted to go.

We loaded our luggage into the car, and we were soon on our way. I sat as still as I could possibly be. I only moved when I had to. I was afraid if I moved the wrong way I would start coughing up more blood. I can't believe I wasn't terrified but, for whatever reason, I felt no fear.

We arrived at my aunt's house and I hadn't bled once. I thought it wasn't going to happen again. But, I was sadly mistaken. Shortly after I got out of the car, I bent over to pet my aunt's cat. Blood started gushing from my mouth. My mother panicked. Within minutes we were back in the car and on the way to see a doctor. I was still as calm as a cucumber but my parents were fit to be tied.

The doctor said my nose was bleeding and running down my throat. He gave me some pills that were supposed to stop the bleeding. They didn't work. My parents sensed something was seriously wrong, so we got back into the car and drove straight home.

About two o'clock in the morning we met our family doctor at the hospital. Though unrelated, his name was Dr. Murphy. I was immediately admitted to the hospital and given a room. It was none too soon. I walked into my room, sat on the side of the bed, and started to choke and cough. The nurse gave me a metal container and within minutes it was overflowing with blood. Dr. Murphy ordered medication, tests, and x-rays. Shortly after I was given several injections the bleeding stopped.

Needless to say, I hadn't been cured. Dr. Murphy referred me to a lung specialist is Springfield. His prognosis was not good. My parents were told, in my presence, I had to have surgery immediately or I wouldn't be alive in three months. I had some weird problem we had never heard of. It was called bronchiectasis. I still don't really know what it is but it's definitely something you don't want.

I underwent surgery that involved the removal of half of my right lung. Two weeks later, I was discharged from the hospital and was sent home to recuperate. That's when I had the most astounding experience of my life.

I was resting in my bedroom. I opened my eyes and saw a bright light at the foot of the bed. It was like a brightly glowing fog. The light illuminated the entire room. As I looked into the brilliant light, I could faintly see the glowing figure of a man. He had a short beard and he was dressed in a white robe. His face was kind and when I looked into his eyes I felt safe and at peace. He looked at me for a moment then extended his arm toward me. I was raised from the bed, as if I floated out of my body, and I stood beside him. I could see my body still lying on the bed.

I can't describe the feeling I had inside. I felt wonderful. Nothing of this earth could ever give anyone that feeling. I felt an absolute feeling of peace. I was consumed by an intense unconditional love that was exploding inside me and radiating from my body. I was in ecstasy. I felt like I was in Heaven.

I wasn't told what was happening, but I knew. I was concerned about the people that I was going to be leaving behind. But I didn't have to say a word. An inner voice told me there was no need to worry. Everyone and everything would be fine.

The man took me by my hand and started walking me toward the light. I was intoxicated with euphoria. After taking only a few steps, the man stopped.

"You have to go back," he said.

"I don't want to go back," I replied.

"You have to go back," he repeated.

"I don't want to go back, I want to go with you," I pleaded.

"You have to go back, you have a destiny to fulfill," were his final words.

It seemed as if we took only a few steps and we were back at the foot of my bed. I again saw my body lying there. I felt sadness inside. I wanted to stay with the man in the light. I didn't want to go back. Before I could plead to stay with him, I was lifted into the air and my spirit was laid back inside my body. The second my spirit's eyes were in place with my body's eyes, they opened, and the experience was over.

I didn't want to come back. I wanted to be with him. But that was not to be. It was over. All I could do was cherish that experience and cry.

I will probably never have that wonderful euphoric feeling again as long as I live. I could have anything I could possibly want a thousand times over and never feel that good inside. I had felt a total unconditional love for everyone and everything. I had experienced something that was not of this world.

I was still crying when my mother came into the room. I instantly told her what had occurred. She never tried to convince me that it was my imagination or a dream. She knew I had a spiritual experience.

Weeks later, my mother told me the doctor hadn't given me good odds of recuperating. The doctor wasn't sure I was going to survive. He thought it was very possible that I might have a relapse. But that was not going to happen. I was well on the road to a complete recovery.

CHAPTER 4

A few months passed before I was up and about. School started, but I wasn't well enough to attend. A tutor came to the house. She was a pretty blonde and very nice. I liked her very much. Her name was Mrs. Bard. I looked forward to the days she came to tutor me, but I did miss my friends at school. I had lots of friends and I received many get-well cards. Every time I opened a card I felt good knowing that someone cared enough to send it. When I received cards from classmates who I never expected to send one, I felt even better.

After my recuperation, I was back in school. The students in my classes were a little ahead of me and I had a difficult time. But I soon caught up. I had completely recovered from my illness, and I was doing well. Soon, my freshman, sophomore, and junior years of high school were over. And, I was going to be a high school senior.

I spent the summer between my junior and senior years, the summer of 1957, in California. My father's parents, James (Jim) and Josephine (Jossie) Newberry had moved to California in 1943. I had grandparents, uncles, aunts, and cousins I barely knew. Some I had never met. I was going to spend three months with them, and I was excited. It was also the first time I traveled by train.

I spent most of the summer in Fontana with my Uncle Jess and Aunt Dorothy. They gave me the grand tour of Southern California, and I finally became acquainted with my father's side of the family. That summer will always be one of the highlights of my life. I loved my father's family and I fell in love with California. I felt as if I was Columbus and, like him, had discovered a New World. From that point in my life, California was constantly in my thoughts.

I wanted to live in California, but the move was not to be in my immediate future. I graduated from Taylorville High School with the class of 1958. Shortly after graduation, instead of moving to California as I had planned, I decided I wanted to be a barber. I soon started attending a barber school in Springfield.

CHAPTER 5

While I was attending barber school I met Judy. She worked for the State of Illinois. We rode to Springfield in the same carpool. Romance blossomed, and shortly after I received my barber license we were married. The wedding took place February 6, 1960, in St. Joseph's Catholic Church, in Nokomis, Illinois. When we first married we lived in Nokomis, and I worked in a barbershop in Pana, Illinois. Our first daughter, Kathy, was born in 1961. Shortly after Kathy's birth, we moved to Taylorville. There I opened my own business. All of my immediate family lived in or near Taylorville, and I was happy to be back in my hometown. Our second daughter, Vicki, was born there in 1965.

I enjoyed living in Taylorville, and business was good. But the nagging desire to move out west was relentless. In 1966 we decided to make the move to California. I sold my business and listed the house with a realtor. It wasn't a good time to sell a house. It wasn't easy for buyers to get loans, and the housing market was extremely soft. I had bills to pay and I couldn't sit around waiting for the house to sell, so I had to find a job.

Rolland Tipsword, who was the Speaker of the House of Representatives for the State Of Illinois, was a friend of mine, so I gave him a call. Two weeks later, I was working in the License Plate Division for the Secretary Of State. There were about fifty employees working in my office, but approximately seven worked in my section. It was at this job that I had a memorable experience.

I can't remember the names of most of the people I worked with. But, the male supervisor, four girls and I were white. One girl, Marlis, was black.

I got along well with other people in the office. I had a good sense of humor and everyone, including the supervisor, liked me. I enjoyed working there, but I noticed every white employee in the section, including myself, got away with almost anything. Marlis wasn't quite as fortunate. The supervisor harassed her constantly. I came to the conclusion that he was not fond of blacks.

Marlis was afraid of the supervisor. She was sensitive and she cried easily but she always kept quiet and endured his abuse. Marlis was a sweet girl and I liked her very much. I didn't like what the supervisor was doing to her.

One day I was talking to Marlis. The supervisor suddenly stood up and yelled.

"Marlis! Shut your mouth and get back to work."

I was the one doing all the talking and he didn't say one word to me. Marlis was humiliated and started to cry. I was tired of watching him abuse her and I was angry. She was afraid to defend herself, and he was making her life at work miserable. I jumped up from my desk, quickly turned to the supervisor, and yelled.

"I've had it! I can't sit here and let this go on any longer. I'm tired of watching you take your hatred out on her. She has done nothing. I was the one doing most of the talking. Why didn't you yell at me? You are prejudiced, and you yell at her because she is too afraid to defend herself or yell back at you. This is going to stop."

The supervisor stood up and screamed back at me.

"You go down to the superintendent's office right now."

I got up from my desk and started to walk out of the office.

"You're damn right, I'm going to the superintendent's office." I fired back at him. "It's time he knows what's going on up here."

To my amazement, all the people in the office started to applaud. It was obvious they didn't like what he had been doing to Marlis either. The supervisor immediately changed his attitude. He didn't really want the superintendent to know what he was doing. He jumped up from his desk and walked toward me.

"Go back to your desk and sit down."

"No!" I replied. "This situation has to stop!"

The supervisor knew he would be in trouble if I went to the superintendent. He became overly apologetic for the injustice he had done to Marlis.

"Let's drop everything and start all over." He said. "I'm sorry for everything that I've done to her. I was wrong. I'll never do anything to hurt her again. There will be no more problems for Marlis."

Everyone in the office witnessed his offer of atonement. I thought he was sincere, so I agreed to drop it. He apologized to

Marlis and I went back to my desk. After things calmed down Marlis thanked me for defending her.

The supervisor kept his word. There were no more problems for Marlis. From that moment on he was after me. Three days later I was transferred to the most miserable job in the department. Filing endless stacks of forms into file folders and filing the folders into cabinets. I hated that job and I was miserable.

I called my friend, the Speaker of the House, and explained the situation to him. Within a few days I had a new job. I was promoted to a position higher than that of my supervisor. He didn't know I had political clout, and when he heard the news he was in a state of shock. Needless to say, he had an immediate change of attitude. His evil ways came to a screeching halt. Now, Marlis wasn't the only one who was happy at work. So was I. Everyone in the office had seen the supervisor sow his evil seed. Now they saw the harvest he reaped.

CHAPTER 6

Two years passed. It was May of 1968. My house hadn't sold and I was still working for the Secretary of State. I drove my car to and from work, and I transported five other people.

One evening, on the way home from work, a man in the group told us about an old black lady who lived in Decatur, Illinois. She was a psychic. Several weeks prior to this, his mother had gone to see her for a reading. His mother told her family what the lady had predicted. He was astounded those predictions were coming to pass. I didn't believe in psychics, but his story had my curiosity aroused. I had to go see Mrs. Joyner for a reading.

The following week, without an appointment, I drove to Decatur for a reading. No one knew when, or if, I was going. I knew only two things about Mrs. Joyner. She was a psychic and she was accurate. I wanted her to know less about me. I made sure no one had a chance to tell her anything.

When I parked the car in front of her house, I didn't know what to expect.

She doesn't know I am coming, I thought. *What if she gets angry because I didn't make an appointment?*

I stepped onto her porch and knocked on the door. The door opened and my knees shook. But the second I saw the expression on Mrs. Joyner's face, I felt at ease.

"I don't have an appointment. But if you have the time, I would like to have a reading," I said.

"Come on in," she remarked with a smile. "I have enough time for you."

I stepped into her house and sat in a chair. She sat in a chair facing me. We exchanged a few words of idle chitchat, but no questions were asked before she started the reading. She didn't even ask for my name.

"I see you driving for some distance to come see me," she said. "You're not from Decatur. What town are you from?"

"Taylorville," I answered.

"I also see you driving some distance to go to work. What town do you work in?"

15

"I work in Springfield," I replied.

She was starting to get my attention and my interest. I didn't expect her to mention things she couldn't have known. She didn't know where I lived or worked. I thought she was making lucky guesses, until I heard her next statement.

"You are going to be moving soon, to a state a long way from here. It is a state with water beside it. This isn't a spur of the moment move. It is something you've wanted to do for quite some time. I also see you signing legal papers. Do you know what they could be for?"

I immediately thought of my plans to move to California, and the fact that my house had been for sale for two years.

"Is it possible it could be for the sale of my house?" I asked.

"That's it," she responded. "Toward the end of June, a young couple is going to look at your house. They are going to buy it. The papers will be signed the first part of July, and you will be moving in the middle of the summer. You will have a chance for a promotion on your job, but you won't take it. You will be quitting your job so you can move. You will not encounter any major problems on the trip to where you are moving, and there will be three people waiting on you when you get there."

I was blown away to hear her say so many things that I knew could happen.

"Are you going to your father's home when you make this move?" she asked.

"No," I replied, "my father passed away when I was four years old."

"Then this man has to be your father's brother because he is like your father."

That statement hit the nail on the head. When I arrived in California I was going to stay with my father's brother and his family.

"Your first job will not last very long," she continued, "but within five years you will never have reason to worry about where your next dollar comes from. Later on, you will get involved in something else and go all the way to the top."

I was flabbergasted. She had no way of knowing any of my plans or what was happening in my life. Yet, everything she said made sense.

I drove home, almost in a state of shock. I was so overwhelmed, I told everyone about my reading. Everyone, except for a couple of people in the carpool, scoffed at me. My wife didn't want to hear about it.

"That's a bunch of crap," she said. "You're crazy, if you believe her."

I had my doubts about that reading, but I knew Mrs. Joyner couldn't have been making lucky guesses. Everything she mentioned coincided with what was really happening in my life. They were things I desperately wanted to happen. But I knew not to make decisions based on her predictions.

On Sunday, June 27, 1968, Judy was out of town visiting her parents. I was home alone when I received a phone call from someone who wanted to see my house. When they arrived I was surprised. They were a young couple. They looked at the house and said they wanted it. I couldn't believe what was happening. It was the end of June, they were a young couple, and they wanted to buy the house. *Could Mrs. Joyner, be right?* I thought.

I called my wife and told her a young couple came to see the house and said they wanted it. She didn't believe me. She thought I was pulling a practical joke on her, because of what Mrs. Joyner had said. I assured her that was not the case, but she still didn't believe me. That is, until she met the couple and she heard them say they were buying the house. Everyone who had scoffed at me about my reading wanted to know how to get an appointment with Mrs. Joyner.

The couple purchased my house and the papers were signed the first week of July. I had a chance for a promotion on my job, but I quit so we could move. The last week of July, we left Illinois. We didn't have any major problems making the move, and there were three people waiting for us when we arrived in California. My uncle, my aunt and one of my cousins were at home. Another cousin was out of town.

CHAPTER 7

Within weeks after arriving in California, Judy and I had jobs and we bought a house in Fontana. Kathy was seven-and-a-half years old, and Vicki was almost three.

My job at Kaiser Steel only lasted for a short time. I was involved in a minor car accident and I was off work for a few weeks. Unfortunately, that was during my probationary period and I was laid off.

I decided to go back to cutting hair and applied for my California license. I passed the state test and soon started working in a barbershop in the Ontario Plaza. I cut hair during the day and attended Chaffey College, in Alta Loma, at night.

In the summer of 1971, I heard about a well-known psychic named Josephine Kellerman. She lived in Yucaipa, California. I was told she was very accurate with her predictions. Due to the accuracy of Mrs. Joyner, I wanted another reading.

Mrs. Kellerman took no appointments. You had to be in front of her house at 6:30 a.m. and wait in your car. She would take no payment. The largest donation she would accept was three dollars.

I went to see Mrs. Kellerman with my Aunt Doralee and Uncle Earl. After waiting four hours it was close to my turn to see her. I was now sitting inside her house. Pictures of Ronald Reagan, Bob Hope, Liberace, and Jim Nabors were on a table in the waiting room. They had signed them for her, indicating she had given them readings.

After waiting for hours, it was finally my turn. When I entered the room, Mrs. Kellerman asked me to be seated in a chair at a small table near her. She was a small, elderly lady with gray hair, and she had an Italian accent. A crystal ball was on the table in front of her. She looked at me, closed her eyes and began to tap on the table top with her fingers. She soon opened her eyes and began to talk. I was immediately in a state of shock. She was telling me everything that was going on in my life. Additionally, she told me I would get a divorce. I did not want to hear that. She then told me she saw me around movie studios and celebrities.

"Later in your life you are going to write something that's going to make you well known and make you lots of money," she added. "And in your later years you will become a spiritual teacher."

I knew no one in Hollywood. I had just moved to California. It sounded far-fetched that I would meet anyone in show business.

When we left Mrs. Kellerman's house I told Doralee and Earl what she had said. I thought it was ridiculous. We all laughed. We had no idea her reading was going to start coming true.

CHAPTER 8

While I was working in the Ontario Plaza, I ate lunch at a restaurant a few doors down from the barbershop. I eventually became acquainted with one of the waitresses in the restaurant. Her name was Brenda Wallis. We became close friends and I soon shared that friendship with her husband, Dick.

Brenda's mother and stepfather, Wanda and Joe DeSantis came into the restaurant quite often. Soon, Wanda, Joe and I were also friends. They met Judy and we started doing a lot of things together.

Joe was an actor. He had been a regular in the "Untouchables" television series and he was in a lot of movies. Joe belonged to the Masquers Club, a show business club, in Hollywood. He invited me to quite a few of their functions. I soon got to know people at the club and Joe encouraged me to become a member. I didn't have to think twice about it.

At the Masquers Club, in 1972, I met Virginia O'Brien. She starred in "The Harvey Girls" with Judy Garland and Cyd Charisse. Plus, she was in other MGM musicals. Virginia and I shared a good sense of humor, and a close friendship blossomed.

In 1972, Kaiser Steel also rehired me. I worked in the Production-Planning Department and continued my night classes at Chaffey College. I also made weekly trips into Hollywood to see my friends at the club.

Shortly after joining the Masquers Club I was asked to model clothes in their fashion shows. On May 5, 1973, I was in a show at the Hollywood Palladium. Jane Withers, a former child star who is best remembered as Josephine the Plumber in the Comet commercials, was the star of the show. That day, Jane and I began a friendship that has lasted for over 30 years. I have also developed friendships with many other people in show business.

In 1975, Judy and I were divorced. That divorce was probably the most miserable experience of my life. As a result, I started therapy with a psychologist in Claremont: Dr. Joe Erickson.

No one will ever hear me say anything unkind about Judy. There were lots of good times during our marriage. I don't choose to remember the times that were otherwise. She has always been a good mother and a good woman. I still respect her very much. She is now remarried and we both share a close relationship with our daughters and their families.

CHAPTER 9

In April of 1976, I purchased a triplex in Fontana. I lived in the back apartment and rented the other two units. The rental income covered my mortgage payment, making it possible for me to live rent-free.

In October of 1976, while living in the triplex, I had another supernatural experience. It was soon after sunrise and I was still in bed. I woke up, opened my eyes and was instantly astounded. LaDonna Higdon, one of my Taylorville High School classmates, was floating face down four feet above my bed. LaDonna was an attractive brunette. She was wearing a blue-gray gown. Her arms were extended perpendicular to her body and she hovered above me like a human cross. She looked at me with eyes filled with kindness and peace. She had a smile on her face.

Within five seconds LaDonna was gone. I couldn't believe what I had seen. I stayed in bed for a few moments and looked around the room. I knew I had really seen LaDonna. It had not been my imagination or a dream.

The last time I had seen LaDonna was in 1968 at our tenth high school class reunion. At that time I still lived in Taylorville. I was on the reunion committee. The Melody Room of the Frisina Hotel was decorated to the nines. "Moments to Remember" was the theme of the reunion.

We had a good turnout for the reunion, and everyone had a ball. I remember talking to LaDonna that night. Like the rest of us, she was excited to see her "old" classmates. I would have never believed the next time I would see LaDonna would be eight years later, hovering above my bed.

When I got out of bed that morning, I couldn't get LaDonna out of my mind. All I could think about was seeing her floating above me, and smiling.

"Why did that happen?" I thought. *"Where was she now?"*

I didn't have the slightest idea where LaDonna lived or who to call and inquire about her. All I knew was her married name, LaDonna Monge.

I decided to call another former classmate, Marilyn Black Malmberg. Marilyn still lived in Taylorville. I knew she would probably know something about LaDonna. It had been several years since I had talked to Marilyn. I didn't know her phone number, so I called her parent's home in Taylorville. By chance, Marilyn was there and answered the phone. After a short conversation I told Marilyn the purpose of my call.

"Marilyn, the strangest thing happened to me this morning. When I awoke and opened my eyes, I saw LaDonna Higdon floating above my bed smiling at me. I can't get her out of my mind. Do you know where she lives? I would like to know how she's doing."

"Oh my God, George," Marilyn replied. "I can't believe this. I just found out a few minutes ago that LaDonna died last night."

Marilyn was shocked to hear why I had called. She knew I couldn't have known LaDonna had passed away. I was equally shocked when I found out LaDonna was dead. I didn't even know she had been ill.

I told everyone about that experience and I wondered why LaDonna came to visit me. I asked myself that question a thousand times. I didn't know that years later I would be given the answer.

I was still having sessions with Dr. Erickson, so I told him about the apparition of LaDonna. I also told him that on several other occasions I had premonitions. Dr. Erickson decided he wanted to do an extra-sensory perception experiment with me. He asked me to hold his watch, close my eyes and concentrate. Then tell him anything that popped into my mind.

The first time we tried it I was a bit apprehensive. I didn't expect anything to happen. He handed me his watch. I held it between the palms of my hands and I closed my eyes.

I sat in silence trying to keep my mind free of any thoughts. Nothing was happening. Then, after two or three minutes of concentrating I began to see something. There were bright colors everywhere. The colors started merging together and formed two objects that resembled two bolts of fabric. The bolts of fabric began to unroll and I could see they were two brightly colored beach towels.

Beach towels? I thought to myself. *How could beach towels have any significance to Dr. Erickson?*

They meant nothing to me and I almost didn't mention them to him. Then, I decided to go ahead and tell him what I had seen. To my amazement, he told me his wife made him go to a store before he went to the office, so he could see some beach towels she wanted to buy. They were made out of a brightly colored fabric.

Over the next few months I continued to tell him things he knew I could not have known. I knew nothing about his home, his personal life, or his family. But I described furniture in his home, told him about his daughters, and told him things he and his family were going to do. On more than one occasion Dr. Erickson was so shocked by what I said, that he flew out of his chair. He was amazed by my accuracy, and so was I. I don't know how it worked, but it did. Since then I have had the same type of experience with other people.

CHAPTER 10

During a session with Dr. Erickson, I told him about a recurring nightmare that haunted me for as long as I could remember. In the nightmare, a man was always chasing me. I would run and hide, but he would always find me. Every time I looked to see who the man was, he never had a face. I always woke myself up, screaming. I was thirty-six years old but I was still terrified every time it happened.

The night before one of my weekly sessions with Dr. Erickson, that nightmare came again. During the session I told Dr. Erickson about it.

"Have you ever thought that you haven't buried your father?" he replied.

Instantaneously, chills went from my head to my toes. Everything began to click. Pictures in my mind began to flash back to the day my father was killed in the car accident. I felt like a bolt of lightning was hitting me. I again saw him covered with blood, lying on the car seat with his face torn off. I suddenly realized the man with no face in my nightmares was my father. For a few minutes I was horrified and distraught. I almost became hysterical. Then my mind began to accept what had happened and I began to feel much better. By the time the session was over I felt fine.

As I was leaving the office, Dr. Erickson said to me. "When you get home, sit down with a pencil and paper and say, 'Who are you?' Then, write down the first thing that pops into your mind. Then say, 'What do you want?' and again write down the first thing that pops into your mind." Although it sounded a little dumb, I promised I would do it as soon as I got home.

I was surprised that I had never thought about the correlation between the faceless man in my nightmares and seeing my father faceless before he died. But that didn't matter, I felt totally at peace when I started the drive home from that session.

Suddenly I felt as if someone or something was in the car with me. That feeling became stronger and stronger and I was becoming more and more frightened. I was ready to pull the car to the side of the road and get out.

"WHO ARE YOU?" I screamed.

I was shocked when I received an immediate response from an inner voice. It was like the voice that I heard when I had the near-death experience.

"I am your father and you are my son," the voice replied.

My heart began to pound. I thought I was going crazy.

"WHY ARE YOU TRYING TO SCARE THE HELL OUT OF ME?" I yelled.

"I haven't been trying to scare you," the voice gently replied. "I've been trying to get through to you all these years. Why do you think you write poetry and music? Don't you remember the poems I sent to your mother when I was in the war? Why do you think all the negative people are taken out of your life? Now you are on the road to where you belong and you don't need me anymore. Now I can be born again."

As suddenly as it started, it was over. But I was terrified. I drove home, went immediately into my apartment, and called Dr. Erickson. When he answered the telephone, I was hysterical.

"I'm going crazy!" I screamed. In between bursts of sobbing, I managed to tell him what had happened.

"Calm down, you are not crazy," he replied. "It wasn't your imagination. It did happen. You're not the first person that has happened to, and you won't be the last. It's happened to lots of people."

With a little more assurance from Dr. Erickson, and with a little time, I began to feel better.

Later that evening I decided to call my mother and tell her what had happened. I was quite surprised at her response.

"Didn't I give you one of the poems your father sent to me when he was in the service?" she asked.

"No," I answered. "I didn't know he wrote any poems."

My mother sent one of his poems to me. When I read it I was quite impressed. My father was stationed at Camp Sibert in Gadsden, Alabama. The poem was written on Camp Sibert stationery and was titled, "Serenade To Camp Sibert." It sounds as if he knew he was going to die. It read as follows.

<u>Serenade To Camp Sibert</u>

*I'm sitting here and thinking of the things I've left
behind.*
*So I had to put on paper, what's running through my
mind.*
*I've dug a million ditches; I've cleaned ten miles of
ground.*
*A neater place, this side of Hell, is waiting to be
found.*
*Although I am no 'gold brick,' I'd like to rest a
spell.*
*When I die I'll go to Heaven, for I've done my hitch
in Hell.*

*We've built a hundred kitchens for the cooks to stew
our beans.*
*We've stood a million guard mounts. We've cleaned
the camp's latrines.*
*We've washed a million mess kits and peeled a million
spuds.*
*We've rolled a million blankets and washed the
captain's duds.*
*The number of parades we've stood in, it's very hard
to tell.*
*But there'll be no parade in Heaven, for we've done
our hitch in Hell.*

*We've killed a million snakes and bugs that look to
us for eats.*
*We've shook a million centipedes out of our dirty
sheets.*
*We've pulled a million briar stickers out of our fatigue
pants.*
*We've battled Camp Sibert skeeters and all those darn
red ants.*
*But when our work is finished, our friends on earth
will tell...*

*How we died and went to Heaven, for we've done our
    hitch in Hell.*

*When our final taps is sounded and we've laid away
    life's cares,*
*We will do a soldier parade up those wide golden
    stairs.*
*The angels will all welcome us and their harps will
    begin to play.*
*Then we'll draw a million canteen checks and spend
    them in one day.*
*It is then we'll hear the Good Lord say, as He greets
    us with a yell,*
*"Take a seat, boys from Camp Sibert, for you've done
    your hitch in Hell."*

Pvt. William S. Newberry 16076307

He added. "I will send you another one as soon as I get it
finished. This isn't very good and all of it isn't so."

I've thought about that incident in the car a thousand times. I
don't know if it was real or my imagination. But for whatever reason,
that nightmare has never returned. Shortly afterward, my sessions
with Dr. Erickson ended. Both Dr. Erickson and I benefited from
them.

CHAPTER 11

In the summer of 1977 came another strange occurrence. I had a dream. At the end of the dream a voice clearly said to me. "Now you've dreamed a complete story. Wake up. Write it down and do something with it."

I instantly awoke. I had to write a spiritual story that would influence people to live better lives. I didn't know where to begin and I racked my brain trying to figure out a good story plot.

I really didn't think about that dream for some time. Then one night, during dinner with Jane Withers, I told her about it. Jane was interested and wanted to hear more. I told her my idea for a story based on that dream. To my surprise, Jane liked my idea. Her enthusiasm gave me the motivation I needed to proceed.

I was ready to start writing when another situation caused me to put the typewriter aside. I was still working at Kaiser Steel in the Production-Planning Department. It was the spring of 1978. It was also the weekend to turn the clocks forward to daylight savings time. I didn't normally work on weekends, but that weekend I had to help take a mill inventory. Unfortunately, I forgot to set my clock ahead and I was an hour late for work.

The job I had to work that day was not in my usual office. I had to work out in the mill. My time card was in my main office so I had to use a temporary card. At the end of the shift, I didn't know what the pay code was for that job, so I asked the foreman. He told me the code to write on my time card. He thought I got paid the same code as everyone else taking inventory. The foreman signed my card.

"If the code is wrong it can be changed," he said.

The next day when I arrived at work, I was immediately told to report to my supervisor's office. I sensed something was wrong but I didn't have an inkling of what was in store for me.

When I entered the office, my supervisor and the superintendent of the production-planning department were seated at a table waiting for me. I sat at the opposite end of the table and faced them. The superintendent, Bill, laid a paper on the table. He held my time card in his hand. The vibrations coming from them were not good. I started to get nervous. I thought I was going to be written up for my being an hour late for work the day before. Bill immediately mentioned the fact that I had been late for work. I was ready to be written up for that when he lowered the boom. Rage and anger covered his face and his voice shook with hatred.

"You altered company documents with the intent to defraud the company. You put the wrong job code on your time card and you didn't think anyone would catch it!" He then pointed to the paper lying on the table.

"Sign that termination paper. You are fired," he yelled.

I was momentarily in a state of shock. How could anyone think I would do a thing like that? I wasn't, and I am still not, perfect. But I wouldn't steal anything, not even a penny, from anyone. My conscience would never let me live with myself.

In an attempt to defend myself I quickly responded. "I didn't know what job code I was supposed to get paid for taking inventory. Stan (the mill foreman) told me to put that code on my card and he signed it knowing it may be wrong. He told me it would be changed if there was a problem."

"Don't sit there and lie!" Bill shouted. "You'll never get me to believe that story!" Again, he pointed to the paper lying on the table.

"Sign it! Now!"

"I will never sign that paper," I replied. "I haven't done anything wrong and I will never sign it."

"I told you that you are fired!" he hatefully responded. "You had better sign it."

I was starting to realize what was happening. Bill had done several mean things to me in the past. I didn't hold grudges and I wasn't a vindictive person, so I had left those episodes in the past. I needed my job. I had to work with him and I didn't want more problems. Now he was showing his true colors. He hated me and he was looking for any excuse he could find to get rid of me. As far as he was concerned, he had the ultimate charge against me. He didn't care about the truth. He didn't care if I was innocent. He was out for blood and he was going to cut my throat. My supervisor, Kathy, was not much nicer than he was. She was helping him with his dirty work and she was ready to swing the ax.

I realized what Bill was doing and I wasn't going to tolerate his evil attitude. "I will never sign your damn paper!" I yelled. "According to you, I'm already fired. What in the hell can you do if I don't sign it? Nothing! You can't fire me twice! I'll never sign anything that's an admission to your lies. You can take that piece of paper and shove it up your ass!"

Fire shot out of his eyes. But I no longer cared what he thought of me. He was out to destroy me and he wasn't going to back off. My efforts to get him to believe the truth were futile. I never signed his paper, but I was still out of a job.

I stayed home and walked the floors for two weeks. Then I received a phone call from the union steward. He was the bearer of good news. The union was fighting for me and a hearing date had been set. My case was going to arbitration.

I went to the hearing, ready to fight. The superintendent was railroading me, and I wasn't taking it lightly. When I walked into the meeting room I was not afraid. Anything that happened to me had to be better than what he had already done.

The second I walked into the meeting room I looked at Bill. His eyes were still filled with hatred. I knew he wasn't going to give in without a fight.

I sat at the table with the grievance committee. Marty Fitzsimmons, the union president, Anne Dunihue, a union representative and personal friend of mine, and the union steward were there to defend me. Bill and Al Gunn, the company arbitrator, sat on the opposite side of the table.

The meeting started with Bill stating his accusations to Al Gunn. When he finished his oration of accusations, I started telling my side of the story.

"If you don't believe what I'm telling you," I said. "I have a witness. Eric (a fellow employee) was present when Stan told me to write that job code on my time card. Eric heard Stan tell me that if the code were wrong, it would be changed. If you talk to Eric, he will tell you that I am telling you the truth. He will tell you the same story."

Bill insisted I was lying. He didn't believe Eric was really a witness. Al set a date for a new meeting and told us to bring Eric.

The second meeting was only a few days later. Eric came to the meeting and backed up my story. It meant nothing to Bill. He accused Eric of lying to protect me. He told Al Gunn that he had spoken to Stan, the mill foreman, about the incident and Stan said I was lying. According to Bill, Stan denied knowing anything about a discussion concerning a job code on my time card.

Al Gunn set a date for another meeting. I wasn't sure what was going to happen to me. All I knew was that Eric and I were telling the truth and Bill was lying.

Bill was a self-proclaimed born-again Christian. He held Bible classes at work during lunch break. Now I knew he was a hypocrite. He was preaching Godliness from his mouth and hiding his evil heart behind the Bible.

The next, and final, meeting was very brief. Again, we sat at the arbitration table. Al Gunn looked me in the eye.

"George," he said. "I spoke with Stan myself. He informed me that Bill never asked him about this matter. Stan told me he never knew Bill was accusing you of falsifying your job code. Stan confirmed that you are telling the truth."

With that, Al looked at Bill. "Bill, it is obvious to me that you are harassing George. I don't need to hear anymore of your lies about what happened."

"Go back to work, George," Al said to me. "You'll receive full back pay. However, I advise you to watch every move you make. I don't think you've had your last problem with Bill."

I could see that Bill was embarrassed. But I was elated. The truth had prevailed but I went back to work only to find out that Al Gunn was right. Bill had been caught lying, he had been humiliated in front of everyone, and he was out to cut my throat. From that day forward, I was never able to relax when I was at work. Bill got some of my immediate supervisors down on me and my work life was miserable.

I was not allowed to take a break and I was not allowed to drive the company truck. Everyone drove the company truck to make deliveries to the mill. I was told that I had to walk. I was harassed constantly. The Labor Board was called on my behalf and Bill lost again.

My co-workers witnessed the abuse I was taking and they begged me to file a lawsuit against Bill and Kaiser Steel. I didn't want more problems, so I didn't do it. That was one of my biggest mistakes. Over twenty years have passed since Kaiser Steel closed its doors and I still get knots in my stomach every time I think of the abuse and mental anguish I endured at the hands of Bill.

I did learn one important lesson from Bill. If a person constantly talks about how good they are or how religious they are, WATCH OUT! They do it because people are never going to see it with their own eyes. Good people never have to praise themselves.

CHAPTER 12

In 1981 Kaiser Steel started having financial problems. I was transferred from the production-planning department to a shipping office out in the mill. My new supervisors were great and my problems at work ended.

I was finally at peace, and I was ready to start working on my book. But once again, I was confronted with problems. I purchased a new black 1981 Buick Riviera. It was loaded with every accessory, plus it had wire spoke wheels and a continental kit on the trunk lid. It was a pretty car, but that was the only good thing about it. From the week it was delivered, I had nothing but trouble with it. During the next two years, the car was in the garage for repairs almost as much as it was in my garage at home. Most of the time I was without transportation. I was a nervous wreck and was too upset to even think about working on my book.

In the spring of 1983 I planned to drive back to Illinois, with my daughter Vicki and her husband Virgil, to visit my parents and family. I took the car to the dealer in March and told them I was going to be making the trip in May, and I wanted the car fixed. The car was out of commission for another two weeks.

I went to pick up the car and I was told the crankshaft had been replaced. A mechanic showed me the old crankshaft. But when I drove the car home, I couldn't tell anything had been done to it. I was no mechanic, but I could tell something was still wrong.

During the next few weeks I made several more trips to the dealer and complained about the car. Each time they kept it for a few more days, and each time I couldn't tell they had done a thing to it.

I knew my car was never going to be repaired by the local dealer, so I took it to another Buick dealership. I didn't tell them about the problems I had been having. To my surprise, I was told the crankshaft was bad and needed to be replaced.

With that bit of news, I went back to the dealership that had supposedly replaced the crankshaft. When I told them what the mechanic at the other dealership told me, I was informed they had already replaced the crankshaft and no more work was going to be done to my car unless I paid for it.

I knew the engine wasn't running right but I didn't have time to argue about it. I had tried to get the car fixed for two months and I was leaving on the trip to Illinois the next week.

The day we left on the trip we didn't drive 20 miles before the engine blew up. Luckily, I was able to drive about 15 miles an hour back to the dealership.

When I found out it would take over a week to fix the car, I was irate. Because of their previous refusal to work on the car, one week of my vacation had been destroyed.

The following week I went back to the dealership to get my car. It was running but something was still wrong. Again, I complained to the man in charge of the repair department.

"The car will be fine to take on your vacation," he said. "If any thing else is wrong with it, we'll fix it when you return."

"But when I return from the trip my car will have over 65,000 miles on it, and it will no longer be under warranty," I replied. "What happens then?"

"Go on your trip and don't worry about it," he said. "We know it's a pre-existing problem and we'll fix it."

That same afternoon Vicki, Virgil and I loaded up the car and, for the second time, headed for Illinois.

We did have a good time on the trip, but the car's engine sputtered all the way to Illinois and back. As soon as I got home I took the car back to the dealership and told them to fix it.

"Your car is out of warranty," the head mechanic said. "We can't work on it unless you pay for it."

"But you told me to take it on my trip and you would still fix it under the warranty when I returned," I responded. "You knew the miles I was going to drive on my trip would negate the warranty."

"Do what you want," he replied. "There is nothing wrong with your car and we aren't going to touch it unless you pay for it."

I couldn't take any more of their attitude and I was at my wit's end. That's when I called General Motors and filed a complaint. A couple of weeks later a factory representative met me at the dealership. I told him every problem I had had with the car and the dealership. To my surprise, even his attitude was lousy.

"There is nothing wrong with this engine," he arrogantly said.

"But another dealership told me the crankshaft was going out before I ever left on my trip," I pleaded.

"There is nothing wrong with the crankshaft and that engine isn't going to be worked on unless you pay for it," he contemptuously replied.

"You may think I'm going to drop this," I angrily responded. "But I'll go to court if I have to."

"I don't care what you do," he replied. "We aren't spending another dime on this car."

By the time I got home I was so angry and upset I was afraid I was going to have a heart attack. After I calmed down, I decided to call the owner of the dealership. I left several messages for him during the next few days. I wasn't surprised when he didn't respond.

I knew I wasn't going to get any help from the dealership so I decided to call the Better Business Bureau. That was the best decision I could have made. I told them what had transpired with the dealership and I gave them copies of repair bills that showed my car had been in the garage for repairs 202 days out of the two years I had owned it. I immediately started getting calls from the dealership and General Motors. Butter wouldn't have melted in their mouths and, though I knew their concerns were phony, they offered to help resolve my problems.

The case went to arbitration. The car was sent to a dealership that had no knowledge of its repair history. Their diagnosis, as expected, revealed the crankshaft needed to be replaced. As a result, a new engine was placed in my car and all the money I had spent on it was refunded. For the first time in two years I had a car I could depend on, but I had gone through two years of living hell and I had accomplished nothing on my book.

CHAPTER 13

In October of 1983, Kaiser Steel Corporation closed its doors for good. I lost my job and my income, but I left Kaiser with assets money can't buy—good memories and friendships. Kaiser Steel is where I met one of my best friends, Orland DeCiccio. He and his wife live about five blocks from me. Orland has always been very supportive of my goal to write this book. Now I had to put my experiences at Kaiser and my car problems in the past, and start writing.

I thought writing a novel was going to be a snap. That was not so. I got writer's block before I finished the first page. I knew what I wanted to say but I didn't know where to begin. I had written poetry and a few short essays, but a book was a horse of a different color. I had to accept the blow to my ego and face the fact that I was not an accomplished novelist.

With high hopes of solving that problem, I joined a writers group; The Pomona Valley Writers Association. It was there that I met Pat Almazan and Edie Boudreau. I acquired two more good friendships that have endured to this day.

I discussed my project with Edie and Pat. Edie volunteered to help me write my book, and Pat's encouragement kept me determined to not give up.

CHAPTER 14

Within a week after Edie volunteered to work with me, I bought a typewriter and a desk, and set up an office in my home. I was ready to start writing.

After collaborating on the book for several months, neither Edie nor I was happy with the fruit of our efforts. The plot was too shallow and most of our text was unacceptable. There was nothing there that would grasp the interest of potential readers. It just didn't have any pizzazz.

I was frustrated, but I vowed I would not give up. I kept writing and rewriting page after page. Most pages went through the rollers of the typewriter directly into the trash. I lost track of how many times I started over. An accomplished writer could have written two novels in the same time it took me to accumulate twenty pages of text. Unfortunately, those twenty pages contained very little commercial material. I had worked very hard and I had accomplished very little. I was getting discouraged. For the time being, I had to put my writing aside.

CHAPTER 15

Money was not plentiful so I rented one of my bedrooms to my friend, Randy Del Turco. Soon after Randy moved in, we joined a group of my Hollywood friends, including Peggy St. Clair, to see a show at The Dorothy Chandler Pavilion in Los Angeles.

Randy recognized a man in the line ahead of us. He was one of Randy's college professors. Soon, all in the group were introduced to Mike and his wife, Paula.

The conversation was centered on Hollywood and celebrity charity events that I was working on with my friend Peggy. Mike and Paula immediately acknowledged they were very interested in getting involved in the Hollywood scene. They wanted to go to the events my Hollywood friends and I were involved with. They wanted to meet celebrities. I was thrilled that in my small way, I could help fulfill their desires. They lived only a few miles from me and they began to call and visit me quite often.

About the same time I met Mike and Paula, I rented out a bedroom to a lady I met through my friend Orland. Her name was Bernadine. Everyone knew her by her nickname, Bernie. Bernie had recently sold her home and was looking for a place to stay until she purchased another house. She had been staying with her niece and her husband. The three of them had problems and Bernie felt it was time for her to move on.

I offered Bernie a room in my home. She moved in right away. Now, all three bedrooms were occupied. The arrangement was good for Randy, Bernie, and me. They were saving money and I needed the extra income.

I continued to work on my book. It still wasn't going well, but I was trying my best to write a good story. I also nurtured the friendships of Mike and Paula, and I enjoyed sharing my home with Randy and Bernie.

Mike and Paula invited me to their home quite often and began giving me expensive gifts. Since we barely knew one another, I was uneasy accepting the gifts. It seemed a bit strange that people who barely knew me were being so generous.

Mike and Paula constantly expressed their desires to go to Hollywood and meet my friends. They had an obsession to meet people in show business. I invited them to be my guests at a banquet for the Southern California Motion Picture Council. It was being held at the Sportsman's Lodge in Studio City. Celebrities were always at those functions and I knew Mike and Paula would enjoy going to meet them.

At that banquet, Mike and Paula were thrilled. I felt good knowing that I was making them happy. I introduced them to my friend, Cesar Romero, and a few other celebrities. We were seated at a table with a few of my Hollywood friends, including the mother of Stevie Wonder, Lula Hardaway.

At that function I mentioned that I wanted to have a party at my home in Alta Loma, and invite my Hollywood friends. Immediately, Mike and Paula volunteered their house for my party. They lived in a large Old Spanish house on an acre of land. My house was very nice but it wasn't big and the back yard was quite small. Mike and Paula insisted that I have the party at their home. I was hosting the party but it didn't matter to me where it was going to be held. I wanted to please them, so I agreed to have the party at their home.

I made the guest list for the party and started sending invitations. As time passed I added a few more friends to the list. I also decided to designate Sybil Brand as the guest of honor and celebrate her eighty-fourth birthday. Sybil's husband, Harry, was once the head of Twentieth Century Fox Studios and Sybil was one of the best known philanthropists in Hollywood.

I kept Mike and Paula informed so they would know whom I had invited, and who had responded. Everything was proceeding quite well, until I received a disturbing phone call from Mike.

"How many people are you inviting to this party?" he asked. From the tone of his voice I could sense something wasn't quite right.

"I haven't really kept track of how many," I answered. "I haven't received a response from everyone yet."

"George," he replied. "You can invite all the celebrities you want. But this is enough of these other people. I thought you were having a party for your celebrity friends."

I was shocked. I couldn't believe my ears. How could they have the nerve to say such a thing?

"I invited all the people I would have invited if I were having the party at my house," I replied.

"Well this is enough of these other people," he retorted. "We thought you were only going to be inviting your Hollywood friends. No more of these other people, OK?"

His statements shocked me. I was at a loss for words. I was disappointed to learn that Mike and Paula were only concerned about having celebrities at their home. I felt a tinge of anger starting to build inside me. I didn't want to create any problems, so I managed to keep it under control. To pacify them, I reluctantly agreed.

I can't remember the exact date of the party, but it was in April on a Sunday afternoon. The weather was beautiful. I think everyone on my guest list attended. Among the guests attending from Hollywood were, Sybil Brand (the guest of honor), Cesar Romero, Lois Nettleton, Jane Withers, Flo Haley (wife of Jack Haley, the tin man in the Wizard Of Oz), Penny Singleton (Blondie), Virginia O'Brien (MGM musical star, The Harvey Girls etc.), Ann Robinson (War Of The Worlds), Chris Costello (daughter of Lou), Lula Hardaway (mother of Stevie Wonder), and my friend Peggy St. Clair.

Everyone had a good time and the party was a smashing success. However, it bothered me that Mike and Paula asked all the celebrities for their addresses and phone numbers. They showed no concern for my local friends. They weren't even friendly to most of them. To a few, they were downright rude. It was obvious that only the people with social status were important to Mike and Paula. I knew I would never invite my friends to their home again.

I went home that evening feeling quite depressed. The friends who came to the party had a good time, but they weren't aware of what was going on behind their backs. I knew Mike and Paula were social climbers. The reason they wanted my party at their house was only to latch onto my Hollywood friends. From that day forward I never felt the same toward them. I managed to maintain a sociable friendship with them but that was all it could ever be.

Months passed. Soon we were into October. November was just around the corner and I decided to invite a few friends to come to my house for Thanksgiving dinner. I included Mike and Paula. When I extended the invitation to them, they asked me whom I was inviting. I told them I was having a small group. I was only inviting a few local friends and a couple of friends from Hollywood. The instant they heard that a couple of friends were coming from Hollywood their immediate response was, "Have the dinner over here."

"I want the dinner at my house," I responded. "I'm only having a small group. I want to have my friends come here."

They weren't happy that I was going to have Thanksgiving dinner at my house. *Why?* I thought to myself. *What difference should it make to them?* Several days later I received a phone call from Paula and my questions were answered.

"Hello, George," she said. "We want to have Thanksgiving dinner here at our house. Invite your friends to come here."

"No," I quickly answered. "I intend to have my friends come to my house."

The next statement from Paula blew me out of my chair. "Hey," she said very sternly. "We want those people for our friends too, you know, and if you can't share your friends we can't be friends."

Fire shot out of every opening in my body. I was furious. I couldn't understand how anyone could be so self-centered and selfish. There was no reason for them to be the way they were. I didn't understand what was going on, or why. I was too naïve. That type of attitude or ego wasn't part of my make-up and I didn't realize those kinds of people existed.

I couldn't believe Paula was being so hateful and rude. But it didn't take long for me to verbally retaliate.

"I want to tell you something right now." I yelled to Paula. "I have friends who are millionaires and I have friends who don't have a penny, and the friends who don't have a penny are just as important to me as the friends with money. You can't treat all of my friends the same, so you won't have my friends or me in your life. I have many good friends to spend my time with. I'm not wasting my time on people like you. You and Mike can take your materialistic, self-centered, evil attitudes and shove them up your asses, and go to hell." I then slammed the phone onto the receiver.

I must admit the last statement I made to them wasn't that nice. But as I have said before, "I have never been and I will never be perfect."

Mike and Paula called my Hollywood friends and invited them to their house for Thanksgiving dinner before I was able to extend my own invitations. Needless to say, I wasn't on their guest list. Neither were any of my "poor" friends. They only invited my friends who were in business, from Hollywood, or who had social status.

On one occasion, Stevie Wonder's mother Lula came to stay a few days at my home. While she was there, Mike and Paula asked me to bring her to their home. I reluctantly accommodated their request. When Lula and I were leaving Mike and Paula's home, I noticed Paula whisper something in Lula's ear. I didn't have to ask Lula what she said. The second we were in the car and the doors were closed, Lula said, "You know what that asshole just said to me? She just told me that the next time I come out to visit not to stay with you, but to stay with them. Those people aren't your friends!"

Shortly after that evening, Lula and I were discussing traveling around the country. Lula had always traveled by plane and had never seen any of the beautiful scenery that exists in America. Two weeks later, we were on a cross country trip, driving from Los Angeles to New York and back. We were gone for over three weeks and we had a ball. Mike and Paula were fit to be tied because I took that trip with Lula. They couldn't handle not being able to control Lula.

I told my friends about the episodes that had taken place with Mike and Paula and I felt confident that Mike and Paula wouldn't get by with their unscrupulous shenanigans. Was I ever wrong.

Mike and Paula lied to my friends and told them I was selfish and didn't want them to be friends with one another. That was the furthest thing from the truth. But for whatever reason, some of my friends didn't listen to me or even seem to care.

Those friends were now the recipients of expensive gifts and they were being wined and dined by Mike and Paula. Of course, they thought Mike and Paula were the nicest people they had ever met. I can't say that I blamed them, because I had fallen for the same bait. They especially wined and dined my friend Peggy because of her contacts to people in Hollywood.

I now knew why they had given me all those gifts. They wanted me to feel obligated to them. I had felt obligated, and I had given them everything they wanted from me. They got their connections to my celebrity friends, and I was no longer needed. I was in their way and I had to be eliminated. Now, they were working on my friends.

When I told my friends the truth about Mike and Paula, some of them didn't believe me, some didn't seem to care and some began to act very strange. Even my closest friends, Orland and Peggy didn't seem the same. I had a difficult time dealing with the hurt I felt inside. I couldn't believe people could be so heartless.

CHAPTER 16

Time flew by. In June of 1986 I sold the house in Alta Loma and bought a large Mediterranean house in Ontario. The house was built in 1928 and it was better suited to house my antique collection. I still needed extra income, so my friends, Bernadine and Randy, moved into the house with me.

It didn't take long to get everything settled in the house. I wasn't working, so I had a lot of time to spend painting, scraping and decorating. I enjoyed working on the house, but it was time for me to face reality. Even with Bernadine and Randy helping make the mortgage payments, my funds weren't going to last forever. I had to go back to work.

I started searching for a job. There were none that paid enough money for me to live on. Good jobs were scarce.

Randy was the district manager for a chain of tuxedo stores. He suggested I open a tuxedo store in Rancho Cucamonga. There was no tuxedo store there, and it was a prime location to open one. I agreed. I had nothing to lose. If the store didn't make it, I would only be right back where I was at that moment, looking for a job. I knew nothing about the formalwear business, but Randy knew it well. He became my teacher and I soon felt secure about opening a store on my own. That November I leased a building, and construction on the interior of the store began.

Also in November, I had a house-warming party and invited approximately 90 of my friends. The house was bustling with people and we were all having a good time. At that party I remember Pat Almazan, my friend from the writers club, saying to me. "Gee, George, you really have a lot of friends."

"No I don't," I spontaneously replied. "I am a friend to a lot of people."

I don't know why that response came so fast. But the more I thought about it, the more I realized it was true.

Nevertheless, it was a fun party. Sybil Brand, Cesar Romero, Virginia O'Brien, Lula Hardaway and several other friends came from Hollywood. Prior to the party I had made the acquaintance of Cindy

45

and Bob Mills, from Chino. I included them on the guest list. That was the beginning of what has become a very good friendship.

I have to laugh when I recall Cindy telling me her first impression of me. Of course, when I met Bob and Cindy I told them that I was going to write a book. I told them about the near-death experience, the apparition of LaDonna and the dream that had inspired the plot.

After we became good friends Cindy said to me, "When we first met and you were telling us all those wild stories that started you writing a book, I thought you were some kind of a nut."

At that time, it was beyond Cindy's wildest imagination that she and Bob would witness most of the unbelievable events that were going to take place in the years to come. Now, they know my story is true and we laugh about their first impression of me.

On December 7, 1986, the grand opening of Tuxedo Junction, Rancho Cucamonga was celebrated. I still remember the comment I made at the opening party. "I hope this isn't the second time a bomb was dropped on December 7th." For those who don't remember, I was referencing the 1941 attack on Pearl Harbor.

My parents and my sister Mary flew from Illinois for the grand opening. My daughters, my California relatives and my closest friends were there as well. They, along with plenty of champagne and food, made the party a smashing success.

To my amazement, that December the store was very busy. By New Year's Eve I had paid all the store expenses for the month and I had made a decent profit.

I've got it made, I thought to myself. *If business is this good and I've just opened the store, within a year I'll be on Easy Street.*

I was on cloud nine. But my excitement was short lived. Here came January. All the December Christmas parties were over and no one was renting tuxedos. I hadn't been open for business long enough to register tuxedos for wedding parties. So, I sat in the store alone. I had no business. I thought I was going to lose everything. I remember sitting in the back room of the store with knots in my stomach and tears in my eyes. I even called Randy and told him I thought opening the store was a big mistake. He told me, "Give it time. You have to be patient. You can't expect a new store to be busy the second you open it." His pep talk made me feel a little better, but I still sat in the back of the store fretting.

Then came February. All of a sudden the store started to get busy. The high schools had formal "King's Balls" and all the boys needed tuxedos. I also started getting brides and grooms coming in to order tuxedos for their weddings. I'll never forget how relieved I felt. The doom and gloom I had endured for a month and a half finally ended. Business was good and I realized I was going to be financially secure.

CHAPTER 17

Business continued to be very good, and I was financially secure. But I wasn't writing my story, and I was extremely depressed over the situation with my friends.

I was soon confronted with the biggest shock of all. I learned Mike and Paula had convinced some of my friends to open a tuxedo store to compete with me, hoping to run me out of business.

Mary Lou, another friend, told me she needed part time work to supplement her income. I gave her a job. Unfortunately, I didn't know she was going to learn the tuxedo business and help Mike and Paula's friends compete with me. She quit with no notice the day the other store opened.

I had been a good friend to Mary Lou. I had helped her several times. I even had her daughter's wedding at my home and I paid for the flowers and limousine. In return, she stabbed me in the back.

I was devastated. I was so depressed I didn't want to see anyone and I had no desire to continue working on my book. I couldn't believe all my new friends were out to destroy me.

One morning when the sun was barely peaking over the horizon, Bernie came downstairs to make a pot of coffee. I had been up all night. When Bernie walked into the kitchen I was sitting on a stool smoking a cigarette. I hate to admit it but at that time, I was smoking three and sometimes four packs of cigarettes a day. Bernie almost had to cut her way through the smoke to get into the room.

"George," she said. "What are you doing up so early? What's the matter? Is something wrong?"

I had never had problems with my friends and I was deeply hurt. I couldn't believe people I cared for could treat me so badly.

"I just can't sleep," I replied to Bernie. "It's driving me crazy just trying to figure out what's happened to my friendships. I don't understand, nor do I believe what's going on. I tell my friends the truth about everything, and they act like they don't believe me. It blows my mind."

"You need to forget about those people, George," she responded. "You don't need those people in your life."

"But Mike and Paula are lying to my friends and getting by with it," I quickly retorted. "Why do they want to destroy my friendships? I've never known people like them before. I didn't realize people like them existed. Neither of them can have a conscience."

"Don't worry about it," Bernie replied. "Everything comes out in the wash. Sooner or later everyone will know the truth."

Bernie didn't seem to be overly concerned about it. But I wasn't surprised. There were times that she hadn't been nice to me. As a matter of fact, she had been downright nasty.

Bernie also constantly stabbed Randy in the back. She had me to the point that I didn't think I could trust him. She wanted me to make Randy move out of the house.

I hadn't given it much thought, but almost every time I was alone with Bernie, she was badmouthing someone. She was constantly in a bad mood. When she would come home from work, it was as if a dark cloud of doom and gloom followed her into the house. She was always getting angry over the pettiest things, and she constantly condemned and judged other people.

If I didn't inform her of every move I made, she was mad. It was impossible to please her. If she wasn't mad about one thing, it was another.

Half the time she wouldn't even speak to me. But if I didn't invite her to go everywhere I went, she was angry. If I made spur of the moment plans and invited her, she was still mad. She accused me of purposely waiting till the last minute to invite her so she wouldn't have time to get ready.

She did what she pleased when she pleased. She would walk out the door and never say where she was going or when she would be back. I didn't care. At least when she was gone there was no tension in the house.

Bernie was being very prophetic when she said; "Everything comes out in the wash. Sooner or later everyone would know the truth." The truth was about to be revealed to me.

One day I received a phone call from Anne Dunihue. Anne had known me longer than any of my friends in California, and she was aware of the heartache I was enduring because of Mike and Paula. I had no idea that everything was about ready to come out in the wash. When I answered the phone, Anne's first statement took me by surprise.

"George," she said, "I've found out the cause of the problems you're having with your friends."

"What do you mean," I asked.

"I know why you have been having so many problems with people disbelieving you," Anne replied. "Your problems aren't just being caused by Mike and Paula."

Anne then dropped the bombshell that blew me away. "Your worst enemy is living in your house."

"What do you mean by that?" I asked.

"George, I just received a call from Bernie and she tried everything she could to get me to turn against you. She was vicious. But I knew she wasn't telling the truth. I told her point-blank that I had known you too long and too well to believe what she was saying."

"What was she saying," I asked.

"She told me you were crazy and needed to see a psychiatrist, and that you were causing all kinds of trouble with your friends. She said you were lying to them, telling lies on them, and blaming them for your problems. She's telling everyone that Mike and Paula were good to you and you are jealous of them."

"When she tried to turn me against you, I knew she was your biggest problem. She doesn't want you to have any friends. She isn't your friend. She's worse than Mike and Paula. She's one of your worst enemies."

I was dumbfounded. Bernie was very moody and hard to get along with, but I didn't expect to hear she was my enemy and the cause of all my problems. But I knew Anne was telling the truth.

Ironically that same evening, before he knew about Anne's call, Randy asked me if Bernie was saying bad things about him.

"Why do you ask me that?" I responded.

"Because, every time you're not around, she always runs you down. She's always trying to get me mad at you. I think she's trying to destroy our friendship."

After a short discussion, we realized Bernie was no friend to either of us. I was uneasy and very disappointed. Even though

Bernie had been nasty to me several times, I always thought she was one of my most loyal friends. Now I knew that wasn't true. Her friendship was a facade.

Randy and I had felt uneasy around Bernie for some time. Dealing with her mood swings had become unbearable. Randy even commented that he felt she was evil. I felt as if the Devil himself was living in my home.

When I confronted Bernie and told her that everything had come out in the wash, she became very angry. Needless to say, so did I. She proceeded to tell me off, and that's when I blew my stack. I ordered her out of my house. Within two days she was gone. She was out of my house and out of my life forever.

Bernie had lied to all of my friends. I thought she was my friend, but she had been in a conspiracy with Mike and Paula.

Bernie lived in my home, so most people who heard her lies thought she was telling the truth. She had even secretly visited and talked to my friend Peggy St. Clair. She had tried to turn Peggy against me too, but she didn't succeed.

Anne and Randy came to my defense. They told everyone the truth about Bernie, Mike and Paula. Within weeks the problems with my friends ended.

Good results came from that experience. I learned a valuable lesson and I realized Mike, Paula, and Bernie had done me a favor. They had taken negative people who were not my true friends out of my life. I didn't lose any true friends.

About a month later, I had a chance meeting with one of Mike and Paula's neighbors. That's when I learned the motive behind their madness. They were social misfits. They had no friends. When they discovered I had several celebrity friends, they wanted to befriend them so they could impress everyone else. They had been showing everyone pictures of my friends taken at my parties claiming they were long-time personal friends of theirs. Mike and Paula didn't want their friends to know the truth so I had to be eliminated.

Mike, Paula nor Bernie had any old friends. When I met them I should have listened to my own philosophy. "If you meet anyone who has no old friends, WATCH OUT! Something is wrong." I didn't listen to my philosophy and I got burned.

Some people are givers and others are takers. Givers are compassionate people who love and care for others. They nurture friendships and are loyal friends. Takers are people who only want others in their lives as long as they can control them, manipulate them, and get what they need from them. When their needs are fulfilled or when they lose control of their "friends," they will walk away from them, never see them again, and feel no remorse. They are only capable of loving themselves.

There are three degrees of friendship. Each degree depends upon the balance between the givers and takers who are involved in each friendship. A 1st degree friendship develops when two takers meet. They may remain acquaintances, but a close relationship will be short term. A true friendship will never develop. A 2nd degree friendship develops when a giver and a taker meet. The friendship will blossom, but it will only last until the giver gets tired of being taken. A taker can never be a true friend to anyone. A 3rd degree friendship develops when two givers meet. They mutually nurture a friendship that will last forever. They share honesty, compassion, loyalty, and a true love for one another. Always remember, we can love many people but we can only be in love with one.

CHAPTER 18

On July 15, 1989, my youngest daughter, Vicki, got married to Virgil Buckner. I bought her wedding gown, furnished all the tuxedos for the wedding, and gave them a reception in my backyard. About 150 people attended. The reception was beautiful and fun. The last people left at 5:30 the next morning. I enjoyed giving them a wedding day that was memorable to them and to me.

In November of that same year, I decided to quit smoking. I knew if I didn't quit, someday a doctor was going to be telling me it was too late. I was finally ready to give it a try. I knew it wasn't going to be easy. I had made several unsuccessful attempts to quit on my own. But this time I tried a new approach. I made an appointment with a hypnotist.

The night before I went for my appointment, I stayed up an hour longer so I could smoke more cigarettes. The next morning I smoked one cigarette after another all the way to his office. Then I stood outside his door and smoked a few more cigarettes before I got the courage to enter his office. I thought to myself, *How am I ever going to quit?* I seriously wanted to stop smoking and I knew if I didn't quit this time, I probably would never stop.

I am proud to say I haven't smoked one cigarette since that day. To my amazement I didn't have a difficult time quitting. I don't know if it was my own willpower, the suggestions from the hypnotist, or both, but I didn't even have withdrawal symptoms.

I also made up my mind I wasn't going to let myself be bothered by someone else smoking a cigarette. I went everywhere I had gone when I smoked. I finally had the willpower to quit.

I also never "bitched" at anyone who smoked. I made a vow that I would never do that. When I smoked, I didn't like being nagged at. It only made me want to smoke more. I was not going to lecture people who still smoked.

Vicki and Virgil's wedding, and quitting smoking, were my highlights of 1989. Vicki and Virgil now have two daughters, Nicole and Marci, and my oldest daughter, Kathy, and her husband, Tom Unsell, have two sons, Bobby and Vincent. And, I am still a non-smoker.

CHAPTER 19

In the summer of 1990, I went with a group of friends on a trip to England and Italy. We toured England, then flew to Venice, Italy for three days. We were gone a little more than two weeks. That was really one of the nicest trips I have ever taken.

When we were in London I was watching the news on television. That's when I first heard there were problems developing in Iraq. I didn't realize those problems were going to eventually place the United States in the middle of a war.

In the first part of 1991, like most other people, I was watching "Desert Storm" on CNN. For some strange reason a poem started coming into my mind. I tried not to think about it, but it wouldn't go away. The more I tried to put it out of my mind, the more persistent the words became. I finally got a pen and proceeded to write my thoughts down on paper. The words kept coming almost as fast as I could write. I wrote this poem in less than one hour.

The Little Blue Man

One person of every race and creed
Was invited by a man,
To join him in a meeting
To bring peace to every land.

With hopes of having Peace on Earth,
From far and wide they came.
Men of all races, men of all creeds,
Some caring, to some just a game.

One by one they arrived and were seated,
Then they all noticed one empty chair.
"Where's our host," yelled one man to the others.
"He's late and I'm sure he sits there."

They started the meeting without him,
But with their differences just couldn't relate.

So, they decided to cancel the meeting
And living in peace had to wait.

Then suddenly their host ran into the room,
Yelling, "Don't leave, I'll help if I can."
But they called him, "Ole Blue," and laughed when they saw,
Their host was a "Little Blue Man."

"You all have the right to be different," he said.
"But your ways forced onto others must cease.
If you lose a debate, but as friends still relate,
You'll help lay the foundation of peace."

"Muslim, Yellow, Christian, White,
Buddhist, Black or Jew.
Your creeds, your colors or beliefs
Aren't why I care for you."

"I am blue, unlike all of you,
Yet, in my house we can all abide.
But a blue man to you is so different,
You're embarrassed with me by your side."

"I'll be on my way, I have friends to meet
Who have differences just like you.
But they live together as "Brothers"
And to them, I'm not, "Ole Blue."

"Many names in the past, you have called me.
I wasn't "Ole Blue" till we met.
But a blue man to you is so different,
The good names you seem to forget."

"When you're "Brothers" I'll come back to see you,
And the same path together we'll trod.
Then call me the names you've replaced with "Ole Blue."
Like Allah, Jehovah, or God."

"If you judge or put down other people,
Change your ways as fast as you can.
For some day you may be condemning someone
Who once was the "Little Blue Man."

I have written several poems. Most of them are spiritual and came to me the same way as this one. Of all my poems, "The Little Blue Man" is my favorite. I hope it helps promote harmony between people of different races and creeds.

No one should have the right to force his or her beliefs onto others. Regardless of race or creed, everyone deserves the same rights. Everyone has the right to their own beliefs, but they don't have the right to force their beliefs onto others. No one has the right to stand in judgment of others. That right only belongs to God.

I'll never forget the time I recited "The Little Blue Man" to a minister. I couldn't believe his reaction. He made no comment and he walked away. He missed the whole point of the poem: Prejudice. Some men are prejudiced and unwilling to accept or love anyone who looks or believes differently than they do. If they didn't know "The Little Blue Man" was God, they wouldn't accept him and love him either. Yet, God unconditionally loves everyone.

The minister was offended because he believed that God only loved Christians and only Christians could ever abide with God. I am a Christian, but I do not believe that. I believe people of all faiths can go to heaven.

None of us knows what the real truth is. Most of our beliefs were taught to us by the religions of our families and friends. If your family and friends are Christians, then chances are that you are a Christian. If your family is Jewish, then you are most likely Jewish. Whatever creed your family may be, chances are, you are the same.

No one can prove who is right and who is wrong. If some of us have been wrongfully taught, it is not our fault. Someday, we may even find out that we are all wrong. Or, as long as we believe in God, that we are all right. I don't believe any church or religion has the right to stand in judgment of another. I believe we will be judged by our deeds just as much as we will be judged by our beliefs.

Sin is in our hearts. If people truly believe dancing is a sin, if they dance they are sinning. If they commit any act they believe is sinful or smiting God, they have committed a sin. I don't believe people will be held accountable for sins they didn't know they were committing. Regardless of religious beliefs, most people instinctively know when they are committing good or evil deeds. We were all born with a conscience. If our conscience tells us we are about to do something wrong, if we do it anyway, we have sinned.

I believe we are all accountable for our deeds. I do not believe people can commit evil acts all their lives and go straight to heaven escaping atonement for their sins. Nor do I believe God would send good spiritual people who worshipped God and lived godly lives, to hell, just because they weren't Christians. I believe people of all religions can go to heaven.

I definitely believe people of all religious beliefs should get their lives in order with God and fight evil. It is the only way to save their souls and bring peace to the world.

Those are my beliefs. If you disagree with me, that is fine. You have the right to your own beliefs.

CHAPTER 20

In July of 1991, I started to think seriously about the story I hadn't finished. It had always been on my mind, and everyone had heard about it, but I hadn't written one word since all the problems had started with Mike, Paula, and Bernie.

I finally stopped procrastinating and became motivated enough to start writing once more. This time I decided that instead of a book, I would write a screenplay. I had never written a script, but I thought it would be easier to write a screenplay than a book.

When I first started to write the script, my neighbor gave me a few pointers. Larry had written a few things and he really was a big help. The best thing was that he kept me motivated.

I was now in the process of developing a fictitious plot. My story had to expose the Devil and his plan to take men's souls and destroy the world. And it had to influence people to fight evil and get their lives in order with God. I didn't know my efforts were going to infuriate the Devil and he was going to strike back at me. I was placing myself in direct conflict with the Ultimate Force of Evil. I had only worked on the screenplay for a month when I experienced the first attack.

In August of 1991, I went on a trip to Asheville, North Carolina. While there, I received a phone call from my store manager. She informed me that she had to fire an employee at the Chino store. She caught the employee stealing money and store inventory. I was a little upset but I had enough employees to cover the stores until I returned.

Two days later I received another call. Another employee had become angry and smashed a computer screen. It was a touch screen, and a used screen to replace it cost a thousand dollars. That news did put a damper on my trip.

When I returned, I took care of the store problems. Then I continued to work on the screenplay. As the story developed, I gained more confidence in my writing. I was happy with the improvements. I finally felt I was going to accomplish my goal and fulfill my destiny. Little did I know the Devil had other plans for me.

CHAPTER 21

On a Sunday afternoon in September, the doorbell rang. I opened the door and was surprised to see my cousin Connie and her boyfriend, Carl. I hadn't seen Connie in at least seven years. I invited them in and we visited for several hours.

The next day they returned. I thought it was a little strange that they were at my house two days in a row, especially since I hadn't seen Connie for such a long time and she only lived 15 miles away.

As we talked, I started to put the pieces together. From comments they made I realized they were homeless. I questioned them about it. They explained they had been dealing with drug problems and they were at rock bottom. They wanted to straighten their lives out and prove to everyone that they were going to change. But they needed help, and they had no one left to help them. They had burned all their bridges.

I didn't pursue the reason, but I thought it was strange that my aunt and uncle wouldn't help them. Even I had second thoughts about getting involved. But I felt sorry for Connie and felt compelled to come to her aid. I didn't think I would sleep at night knowing she was living on the streets. They seemed to be sincere about changing their lives, and I thought it might be the one time that someone's help would make a difference.

I didn't know anything about Carl, but I had always been fond of Connie. She had never been a bad person. She was not lazy, she was always nice to other people, and she always managed to smile even when things weren't going well for her.

From statements Connie had made, I felt there was a chance she would free herself from the bondage of drugs. I knew she would never win the battle as long as she was living on the streets, so I allowed Connie and Carl to move into my house.

A part of my story dealt with drugs and how Satan uses them to take control of people's lives. Now I was seeing an example of that happening in my own family.

Drug dealers, as well as gangbangers and other evil people, are the Devil's Disciples. They help Satan influence weak people to live evil lives. He wants to take peoples' souls and gain control of the world. The Devil's Disciples, or False Prophets, are not just the Hitlers, Mussolinis, Titos or Bin Ladens of the world, they are also ordinary people that we meet every day of our lives. They are people who hide their evil intentions behind facades of kindness, love and friendship. Anyone who encourages another person to think or do evil is, at that moment, a Devil's Disciple. Most of them aren't even aware that Satan is using them.

To have influence over his victim or victims, the Devil's Disciple must become a False Prophet and gain his victim's trust. With that trust the victim is more vulnerable, and the will of the False Prophet becomes easier to accomplish.

Weak people are easy prey for Satan. His disciples befriend them and gradually induce them to live evil lives. False Prophets thrive in all walks of life. They are not restricted to any social class. They could be gangbangers, friends, high-ranking politicians, or even respected ministers.

Another example of a False Prophet is a drug dealer. Drugs are the Devil's keys to people's souls. Once the dealer has influenced his victim to try drugs, he has placed the key into the lock that protects that person's soul. His victims become addicted to the drugs and they gradually lose their consciences. And gradually, Satan takes control of their lives.

As a victim's conscience dwindles away, he feels less and less guilt for committing ungodly deeds. Once his conscience is gone, the Devil is in control of his will. Nothing is sacred to him. He can lie, cheat, steal, and even commit murder, yet feel no remorse. He can beguile his friends and his family just as easily as he can hoodwink a total stranger. When his conscience is totally gone, he has no desire to free himself from the bondage of evil and regain his soul. His soul is lost. Even if he believes God and Satan do exist, he no longer cares. He feels no remorse for his evil deeds and he doesn't believe, or care, that he is an evil person. Now, it takes nothing short of a miracle for him to attain the willpower to amend his ways and regain his soul.

Not unlike Osama Bin Laden and other terrorist leaders, many false Prophets profess to be doing God's will. They brainwash their followers and gain their trust. Once their followers believe they are working for God, they are easily manipulated into doing anything.

False Prophets may even tell their victims, "God told me to give all my possessions away. I want nothing from you. I will never ask you for anything."

They know they will not have to ask. They are being devious. They will later manipulate their followers to give them anything they want. They will eventually control their followers and all their possessions. Many phony ministers quote and use the word of God to manipulate their victims into sending them money and material possessions.

The most dangerous of all False Prophets are the likes of Osama Bin Laden and other Islamic radicals. Hatred, evil, and the greed for power possess them. They convince their weak followers to murder, steal, terrorize people, attack countries and promote war, all in the name of God. They and their followers are convinced God sanctions their actions, and they feel no guilt for their evil deeds. They have been brainwashed by the forces of evil to believe they are the servants of God. They even commit suicide for their cause, falsely believing they are sacrificing their lives for God and will spend an eternity with Him in paradise.

There is no such thing as a "Holy War." God does not give the followers of any religion the right to commit murder and destroy other religions or countries in His name. But, victims of evil aggressors do have the right to fight back in defense of their lives and their countries. False Prophets and their followers, who promote evil in the name of God, must be stopped. We must fight them until they are defeated or they will destroy the world.

CHAPTER 22

Shortly after Carl and Connie moved into my home, Connie told me they were in dire need of money to pay off their debts. With the promise of getting paid back when Carl found work, I loaned them the money they needed. I was also told that Carl could repay me from an insurance settlement he was going to receive.

Carl regularly borrowed my car or my van to look for work. Within a week he found a part-time job to help sustain them until he was able to find full-time work. My van became his transportation to and from work. For the next few weeks he was gone quite a bit.

Then, one day I received a phone call from Carl. "George, I'm in Fontana," he said. "I don't know what happened, but the wheel fell off the van. You have to send a tow truck or come after me."

I didn't understand how a wheel could loosen itself and fall off, so I drove to Fontana to see it. When I inspected the damage I wasn't surprised to see that the wheel hadn't fallen off the van without reason. Carl had had a collision and wouldn't admit it. The lug bolts were broken off and the wheel was bent. It was impossible to get it fixed easily, so I had the van towed to my house.

Carl said he had been driving the van and the bolts broke off the wheel by themselves. But I knew that wasn't the truth. It was obvious he had hit something very hard. But no matter what had happened, I knew he would never tell the truth, so I didn't pursue the discussion any further. I had nothing to gain.

When I took the van in for repairs, I was told that the entire frame of the van was bent. He had hit something going at a high rate of speed, and the force of the impact had broken the lug nuts and caused the damage. I spent over a thousand dollars on repairs. The van has never driven the same as it did before that accident.

For the next month and a half my problems continued to mount. For fifteen years I had owned the apartments. They were a steady source of income and I had never had major problems. Now, drug dealers and gangs were moving into the neighborhood and my long-time tenants were afraid. Two of them moved out within a month. I lost the rental income and I had to renovate the apartments before I could rent them out again. I lost thousands of dollars and lots of sleep.

In the meantime, my suspicions that Carl was still involved with drugs were confirmed. He had not been using my van and my car to go to work; he had been driving them to make drug deals.

I realized Carl was a hopeless case, but I retained my confidence in Connie. As long as she was sincere about changing her life for the better, I still wanted to help her. Even I had tried "speed." But only trying it was enough for me. I saw what it was doing to people and I didn't want anymore to do with it. I thank God I didn't get addicted.

I knew Connie and Carl's lives were not going to get any better as long as they were abusing drugs. I was trying to help them get their lives in order, and they weren't even trying to help themselves. I couldn't deal with their drug problem any longer. So, I ordered them not to have any drugs around my house. That's when I discovered they didn't care what I wanted. The drugs were in control of them, and they weren't going to change. Helping them was wasting my time and my money, and I wasn't able to work that much on my screenplay.

I soon noticed that some of my possessions were missing. In addition, about a thousand dollars in change disappeared from two slot machines I had in my bar room. I kept the money in the machines so friends could play with them. They used my money and left it in the machines. That way no one was using the machines for gambling.

I confronted Connie about the missing money, but she denied having anything to do with it. I couldn't prove who took the money, but I knew the guilty person had to be living in my house.

Before I found out who took the money from the slot machines, I was confronted by a more devastating problem. I came home from work to find a friend of Carl's using my computer. I let him know I didn't want other people to use it. He assured me he knew all about computers and there would be no problem. But the next time I went to work on my script, everything had been wiped out. He had destroyed everything I had written. On top of that, he had destroyed my program disks and I couldn't get the programs loaded back into the computer. I was both heartbroken and furious. Everything I had written was gone. Plus, I no longer had a computer that worked. Eventually, I ended up giving it away.

I told Carl I could no longer handle him or his drug problem. He had to move out of my house. Connie had always said that if Carl didn't stay away from drugs he could go on his own. So, I told her that I would still help her if she wanted to stay with me.

"Connie," I said, "If you want to change your life, I will get you a makeover. I'll help you get a decent job, and I will introduce you to a nicer group of people but Carl has to go. He doesn't want to change his life; he likes the way he lives."

"I'm sorry," Connie responded. "If Carl goes, I go with him. I can't live without him."

That was the last thing I had expected to hear from Connie. But that was her decision. The next day they both moved out of my house.

CHAPTER 23

I don't regret trying to help Connie and Carl. I felt I had done the right thing. I had every intention of being a Good Samaritan. I truly thought I was going to help free them from the clutches of the Devil. But I didn't know drugs had already consumed their consciences and the Devil had more influence on them than I did. I had become another one of their victims. Like everyone else who had tried to help them, I had been used.

I knew there was nothing more I could do for them and it was time for me to go forward with my own life. But I did feel disappointed and defeated and was ashamed to tell people I had failed. I didn't want to hear all of the "I told you so's" from the people who had warned me not to help them.

When Carl and Connie moved into my house, Carl had offered to help remodel my kitchen. It was his unsolicited gesture to show his appreciation for helping them out. Needless to say, he never helped work on the kitchen. As a matter of fact, he never once helped with anything around the house.

Connie was the opposite of Carl. She wasn't lazy and she was always there to lend a helping hand. We spent a lot of time working together on the kitchen. But when they moved out of my house, the cabinets were only partially stripped and the kitchen was torn apart. Carl wasn't going to be keeping his promise. I wasn't a carpenter, and I didn't have the money to pay anyone to fix my kitchen. I had been burned again.

A few days after they moved out of my house, I discovered they had found my credit cards and had run up charges on them. They had also forged checks on my store account and Carl had stolen store inventory that had been stored in my garage. He had been both selling it and giving it away. I couldn't believe the whole time I was handing them money and helping them, they were ripping me off. Again, I lost thousands of dollars.

I can't describe the feelings that churned inside me. They were a combination of anxiety, depression, worry and pain. I don't think there was one bad emotion I didn't feel. But this time I didn't just sit there and feel sorry for myself. I did something about it.

I called the Ontario Police Department and filed a police report. I told them everything. I showed them the forged credit card receipts and the forged checks. I had all the proof I needed to prosecute both Carl and Connie. But I couldn't do it.

Connie's mother and father were good people and they had been very good to me. I couldn't do anything that would cause them more pain. Their son Doug had been caught manufacturing drugs and he was doing time in prison. Now, their daughter was a drug addict. Drugs had shattered both of their children's lives, and they were the innocent victims of Satan's work. They had suffered enough. To spare them grief, I decided not to press charges against Carl and Connie.

Instead of wallowing in my grief, I started working full time on my screenplay.

CHAPTER 24

Soon, Christmas and New Year's Day were past, and it was January 3, 1992, a day I will never forget. I was home alone working on my script when the doorbell rang. I opened the door and was startled by six or eight sheriff's deputies who barged into the house. They were not friendly. One of them handed me a search warrant and ordered me to go into the living room and sit down. I was shaken by their abruptness and I was speechless. I laid the search warrant on a table and went into the living room. Intimidated, scared and shaken, I sat on an antique sofa facing the fireplace.

One of the deputies stood with his back to the fireplace and kept his eye on me while the other deputies proceeded to search my house. I knew no drugs or drug-related paraphernalia were in or around my house. That is unless they had been stashed there without my knowledge. One of my dogs jumped into my lap. I held him close and for a brief moment I prayed silently. I couldn't believe this was really happening to me. It was a living nightmare.

After I had time to calm down and organize my thoughts, I realized the real reason they were there. They were on Carl's trail.

At first I was afraid to say anything. Then I held a tight grip on the dog and got the courage to speak up.

"I know who you're looking for," I uttered. "My cousin and her boyfriend have a drug problem and they were living with me. When I let them move in, I knew they had problems but they told me they wanted to get off drugs and change their lives. I believed they were sincere. They had no one else to turn to, and I thought they'd change if I helped them. I gave them a place to live, I fed them, and I even gave them money. But the whole time I was trying to help them, they were ripping me off. They stole from me, forged checks on my store account, and ran up charges on my credit cards. I finally realized they were using me and they weren't going to straighten out their lives, so I made them get out of my house. That same day, I called the local police department and filed a police report. I even called a second time and asked them to bring dogs to search my house. I wanted to make sure there were no drugs hidden anywhere."

Those statements got the attention of the deputy in charge. He asked one of the other deputies to call the local police department to see if my story could be corroborated. Evidently it was. The search was discontinued, and my house was not torn apart.

One of the deputies saw my screenplay lying on a table. At that time, the title of it was "False Prophets." The deputy picked it up and browsed through it. "What's this?" he asked. His question got the attention of several other deputies. I briefly told them what it was and what it was about.

Those deputies probably thought my story and I were both big jokes. I was writing a screenplay that was a condemnation of Satan and drugs, and they were searching my house for materials with which to manufacture methamphetamine.

Despite what they thought, they did have a noticeable change in their attitudes. They were much nicer. They even told me a few stories about things that had happened to victims of drug addicts and drug dealers. I felt fortunate that the deplorable stories they told had not happened to me.

After the deputies left, I remained sitting on the sofa in a daze. It was hard to conceive that I hadn't had a bad dream. But it wasn't a nightmare. It was real. I felt nauseous, humiliated and depressed. *What would my neighbors think?* I thought. *And my kids? And my family?* I could have lost my home if drugs had been found in it. I was depressed, but I felt fortunate that I had been wise enough to file a police report.

For a few moments I sat in a stupor and stared into space. Then the search warrant popped into my mind. I got up from the sofa and walked to the table where I had put it. I picked it up, walked back to the sofa, sat down and started reading. I had no idea I was in store for one of the biggest shocks of my life.

The court issued the warrant to give the deputies permission to search my home and my personal property. They were to search for chemicals or paraphernalia with which methamphetamine could be manufactured. My home was the first thing they had been given permission to search. I was the second and my cars were the third. A long list of chemicals and drug-related paraphernalia the deputies were to search for was also included on the warrant.

I read the detailed description of my house and property. A description of me was next. Then, I started to read the description of the cars they were going to search. The first car listed was my 1981 Buick, California license: 1CRE851. The second car listed was my 1988 Dodge, California license: 2KNB997.

I wondered for a moment why a third car was listed on the warrant. Then I started to read its description. It was a 1972 Ford. *Oops,* I thought. *That car isn't mine. Someone made a mistake.* Then my eyes saw the California license plate to which that car was registered. I couldn't believe what I read.

"Oh my God," I yelled, as I jumped to my feet. My heart pounded, and a bewildering heaviness inundated my entire body. My blood rushed from my head to my toes and my skin burned and tingled.

When I regained control of my senses, I took another look at the search warrant. I thought that perhaps I had misread the license plate registered to the Ford. But I hadn't. It was still there: 1972 Ford, California license, "LUCIFUR." I shook my head in disbelief. I had a hard time accepting that something like that could happen. (A copy of that warrant is included with other documentation at the back of this book.)

How did that car with that license plate get on the search warrant, I thought. *This is uncanny. My story is assaulting the Devil and nothing but bad has happened to me.* The name "LUCIFUR" on that warrant triggered an apprehension that perhaps I was being pursued by something evil.

When I examined the rest of the search warrant I noticed that Judge Ellen Brodie accidentally wrote the wrong date on it. She missed the recent change of the year and wrote the date as being January 3rd, 1991. The real date was January 3rd, 1992. The wrong date, as well as "LUCIFUR" being spelled with a "U," instead of an "E" was meant to happen. Later, I will explain why.

That same afternoon I called the deputy in charge of the case. His name was Richard Hahn. When he answered the phone, he had no problem remembering who I was.

"I hate to bother you," I said. "But I have to ask you a question."

"Sure," he responded. "What do you need to know?"

"Where did you get this 1972 Ford with this "LUCIFUR" license plate on this search warrant?" I queried. "I only have two cars. I've never heard of this 1972 Ford, and I would never have the name "Lucifur" on anything of mine."

"We ran a DMV report for any car registered to you at your address," he replied. "When we received the report all three of those cars were on it."

"But how can that be," I asked. "I've never owned such a car, and I would never want to have anything with the name Lucifur on it."

"We only asked for the report," he said. "You'll have to check it out with the Department of Motor Vehicles."

That I did. There was no such car registered to me. But there had been a 1972 Ford with the "LUCIFUR" license plate registered to another George Newberry. In 1982 the title of that car had been transferred to someone in Compton.

I was baffled by the mysterious appearance of the name "LUCIFUR" on that warrant. When I showed it to my friends, even they were astonished. I kept the warrant for a short time, but too many bad memories were associated with it. I had to get rid of it. I threw it in the trash and hoped I could forget what had happened.

CHAPTER 25

I got rid of the warrant but I didn't get rid of the memories. *How could such a thing happen?* I kept thinking. *Of all things that could have been on that warrant, why did it have to be "LUCIFUR?"* The more I thought about it, the more I realized it was meant to happen.

Millions of personalized license plates are issued in the State of California. But no other name would have had the same impact on me as "LUCIFUR." I suppose some people may have written it off as a computer glitch or a coincidence. At first, I tried to convince myself to believe that too. But I no longer believe there is any such thing as a coincidence. Everything happens for a reason. Everything printed or written on that search warrant was meant to be exactly as it was.

Time passed and my problems continued to get worse. I was bombarded with a relentless barrage of bad news and financial losses. Every week I was losing thousands of dollars. I finally got to the point that I couldn't cover the losses any longer. My money had run out and I had to take out a second loan on the apartments.

During this time period I received a surprising phone call. It was a voice from the past. Lenny was an old family friend and it had been at least fifteen years since I had seen him. His sister Patty had been married to my cousin, Connie's brother Doug.

Lenny had heard about the episode with Connie and Carl. He also heard that my kitchen was in need of repair. Lenny was a good carpenter and he was temporarily out of work. So, he offered to repair my kitchen for less than it would normally cost. The kitchen definitely needed repair, and I definitely needed to save money. Lenny started working on the kitchen the next day.

Lenny and I worked well together. While he tore out walls and remodeled the kitchen I worked on the cabinets and kept things clean. He was a pleasant person, and he had a good sense of humor. While we worked we shared a lot of laughs, but we talked about serious things too. I told him all my problems and he told me his. Lenny had dealt with his own drug problems. Now they were a part of his past. He was straightening out his life. He had goals for his future and he was planning on getting married. I was happy to see that, unlike Connie and Carl, Lenny had whipped his problems and was well on his way to recovery.

During the daytime Lenny and I kept working on the kitchen. At night I kept working on the screenplay. Day and night, the Devil kept working on me.

The apartments were still a dilemma. Between problems with tenants and more monetary losses I was going crazy. There was no way I could relax. I never had a stress-free moment.

CHAPTER 26

In February of 1992, my parents flew in from Illinois to visit me. As much as I loved them and as happy as I was to see them, I'm sorry to say my usual bubbly personality was gone.

I didn't want them to know anything was wrong. There was nothing to gain by telling them all the bad things that had been happening. They had flown two thousand miles and they were thrilled to see me. I loved them and I was happy they had come, but I had a hard time keeping a happy face, and they could tell something was wrong.

"What's the matter with you," my mother constantly asked. "I've never seen you act this way before. What's wrong? Are we in your way?"

"There's nothing wrong, Mom," I always replied. "Just drop it. There's nothing wrong!"

My mother was no dummy. She knew something was wrong, and she wasn't going to drop it until she found out the truth. I was determined that was not going to happen. I was trying my best to endure the agony of a financial and emotional crisis, but I couldn't let them know that. I felt if I did, they would go back to Illinois and worry themselves sick.

My mother would not give up, and her persistence was making me even more distraught. I didn't want to keep denying there was anything wrong, so I started making excuses to stay away. That was a big mistake on my part, and it was the straw that broke the camel's back. One day I left them alone for several hours. When I returned home, they were gone.

I went into the kitchen where Lenny was working. "Where are my parents," I asked.

"They're gone," Lenny replied. "They said they felt like they were intruding on you, and they went to stay with your daughter. I tried to explain to them that you had been dealing with a lot of stress, but that made no difference. They don't think you want them here, and they're going to go back home."

That was the last thing I had wanted them to think. A jolt of anguish ripped through my body. I had hurt two of the most important people in my life, and it was my own fault. I had fallen into the Devil's trap. Instead of being honest with my parents from the beginning, I had tried to hide the truth. As a result, I had unwittingly devastated them.

I got in my car and drove straight to my daughter Vicki's house in Fontana. I had to face the fact that I had caused this problem, and I had to resolve it. It was time for me to tell the truth.

By the time I arrived at Vicki's house, I felt like a first-class heel. When I stepped into her living room and faced my parents, my self-esteem couldn't have been lower. I was ashamed to look them in the face. The way I had dealt with my problems was wrong. I had abused my parents and I had to make things right.

"I'm so sorry." I said to them. "I would never do anything to hurt either of you. I've been a nervous wreck trying to hide the truth. I didn't want you to know the bad things that have been happening to me. I didn't want you to be upset or to worry, but I've accomplished the very thing I didn't want." I had to make things right, and the only way I could do that was to tell the truth.

They listened intently as I told them everything that had been going on in my life. I thought it would be hard for them to believe some of the things I was telling them, but they seemed to know I was telling the truth. Their main concern was that I still loved them and they were not unwelcome guests in my home. They had never been angry with me; they had been hurt. They knew I loved them very much, and I was sorry for causing them pain. Being the loving parents they were, they went back to my house and all was forgiven. During the balance of their visit we had a good time and they flew back to Illinois feeling much better. I still regret I caused them so much grief during that visit.

CHAPTER 27

My friends couldn't believe so many bad things were happening to me. Every time I saw Orland, Bob or Cindy the first thing they would say was, "Have any more bad things happened yet?" Unfortunately, most of the time I had to respond, "Yes."

As if I weren't having enough bad luck, my car broke down. My van was still setting in the driveway with a bent frame, so I had no transportation. The husband of one of my employees was a mechanic, so I took the car to him for repair. For transportation, I had to rent a car.

During the next few weeks, the mechanic called several times to tell me the car was repaired and I could pick it up. But each time, before I could leave for his shop, he called back and told me not to come after it.

"This is the weirdest thing I've ever seen," he said. "Every time I think I have the problem fixed, as soon as I call you to come pick it up, the damn thing starts cutting out again. It's driving me crazy!"

"How much longer do you think it will take to fix it," I asked. "I'm wasting a lot of money renting a car."

He assured me he was doing his best to find the problem, and I would have my car back as soon as possible.

Soon, my car had been in the shop for six weeks. The bill for the rental car was starting to add up. My car had to be fixed before long or I had to make other arrangements for transportation.

That afternoon, I was working in the kitchen with Lenny when my friend Orland stopped to visit. I told him about the problems the mechanic was having with the car. When Orland heard I had been renting a car for six weeks, he almost fainted.

"Take that rental car back," he said. "I can get along fine with my car. Use my truck until your car is fixed. Don't waste any more money on a car rental."

The following day I returned the rental car to Hertz. I checked the car back in and all was fine, that is, until I was handed the bill. When I saw how much I was being charged, I almost died. I can't remember exactly how much the bill was, but only half of the charges were for the car rental. The other half was for insurance.

Hertz had billed me for a six-week insurance premium that would have paid my car insurance premium for a full year. I wasn't that concerned about it because I had declined the insurance when I rented the car. I had told the girl I did not want the insurance. At that time, she handed me the contract and told me to initial it to show that I declined the insurance. With the intention of not accepting the insurance, I initialed the contract on the line next to the word "decline." Now I was being told that I had accepted the insurance, so I asked the girl to show me the contract.

She handed the contract to me and I carefully read the entire statement. The section that I had initialed said, "I accept or decline insurance as stated above." I had placed my initial on the line next to the word, "declined." In the section above, it stated that I accepted the insurance. My initial next to the word "declined" didn't mean I had declined anything.

I was furious. There was no point to have me initial anything in that section. In my opinion, that part of the contract was purposely written to be misleading and I had been tricked. I wondered how many other people had put their initial next to the word declined, thinking they had declined the insurance, and got stuck with a bill like I did.

I refused to pay the insurance, but Hertz charged it to my American Express Card. I told Hertz that as far as I was concerned, I had been duped by a poorly written, misleading contract. I was not going to pay the insurance premium, and I was taking them to court. I called American Express and had the charges withheld pending a settlement. After several weeks of arguing, Hertz agreed to deduct fifty percent of the insurance cost. To avoid a court battle and to end the matter, I agreed to pay it.

I still think that contract was purposely written to be misleading, and I still wonder how many other people were tricked by that clause.

The day after Orland loaned me his truck; I drove it to the store. When I was ready to return home, I got in the truck and started the engine. For a moment I was distracted by something that was lying on the seat. I wasn't accustomed to a stick shift and I didn't realize I was holding the clutch to the floorboard, putting the truck out of gear. The truck slowly rolled backwards and hit the car behind me. It left a dent about the size of a pea in the car's hood. The truck didn't suffer a scratch. I waited for the owner of the other car to return before I left. I explained what had happened and I gave him my name, phone number and address.

When I returned home, Lenny was working in the kitchen. When I told him what had happened, he almost went berserk.

"What's going to happen to you next," he exclaimed. "Are you kidding me? This is unreal."

It was unreal. Too many bad things were happening. I didn't have time to recuperate from one misfortune until I encountered another.

I didn't want to involve Orland or his insurance company in the accident, so to keep from having a problem I paid four hundred and seventy five dollars cash to the owner of the car. I had suffered another financial loss, but at least the case was closed and the owner of the other car was happy. The dent was so small, he probably pocketed the money and never got it repaired.

CHAPTER 28

A couple of days later, Lenny and I were working in the kitchen. I was staining cabinets and I was getting stain on my rings.

"Give me your rings so I can put them in the cabinet." Lenny said. "You're getting stain all over them."

I didn't care if stain got on the rings, I figured I could clean them. But I took off three rings that I was wearing and handed them to Lenny. One ring had a three-carat center diamond, one had an eighty-point diamond and the third had multiple channel set diamonds. Lenny placed the rings in the cabinet and we continued our work in the kitchen.

The following day I saw the rings still intact setting on the shelf. The day after that, Lenny and I were again working in the kitchen. When I opened the cabinet where the rings were stored I noticed the eighty-point diamond was missing out of the mounting. My heart almost stopped beating, and I felt weak in the knees.

"Oh my God!" I said, as I turned to Lenny.

"What's the matter, now?" he replied.

"The diamond is missing out of my ring," I yelled. "I know it was there yesterday. I saw it."

"This is too much," Lenny said, as he looked at the empty mounting. "This is unreal. You must have really pissed the Devil off. Maybe you should think about dumping that story."

"No," I exclaimed. "That's exactly what the Devil wants me to do."

My work in the kitchen instantly came to a screeching halt. I spent the rest of the day inspecting every inch of everything looking for my diamond. I even checked the yard within ten feet of the kitchen door to make sure it didn't get swept outside. It was nowhere to be found.

Lenny suggested that the stone might have been missing when I handed him the ring. I knew that wasn't so. I had seen the diamond in the ring when I looked on the shelf the day before. A diamond is only a material thing, but I was quite despondent over its loss. That diamond meant a lot to me. It was a good stone and was worth a few thousand dollars. But its sentimental value made it priceless. I had bought it when I lived in Illinois. I had scratched with the chickens to pay for it, and I had owned it for almost thirty years. The memories associated with that stone couldn't be replaced.

Several days later, I took the empty mounting to a local jeweler. After checking the mounting, he told me the prongs that held the diamond in place had been pried open. Someone had stolen the diamond. That bit of news made me feel even worse.

How could someone come into my home and do that to me, I thought. *I don't deserve the bad things that are happening to me? Why are they so relentless? Aren't they ever going to stop?* I really knew why so many bad things were happening to me, but I didn't want to give the Devil his due.

I had nothing to gain by pursuing the matter any further. Doing so would only make matters worse. I knew I would never know who took the diamond, and I would never get it back. I thought it was best to chalk it up as another loss and drop it.

CHAPTER 29

From that day forward, the topic of good versus evil dominated the conversation between Lenny and I. I kept encouraging him to be strong, to fight temptation, and to stay on the right track. I continuously gave him spiritual pep talks.

"What goes around, comes around," I said to Lenny. "The seed that you sow is the harvest you reap. Do good deeds, be honest and live a good life no matter what happens to you. Then good things will come back to you. But always remember the good things you do have to come from your heart. If your seed is sown only because of the expectancy of receiving good things back, then your seeds will never reap a harvest. Those seeds were never sown; they were put up for trade. To give, you can't expect to receive anything in return."

"Then why are all these bad things happening to you?" Lenny asked. "You're not a bad person."

That statement caught me off guard. That same question had crossed my mind more than once. I couldn't understand why all the bad things were happening to me. I didn't feel I had done anything in my life that would cause me to deserve such bad Karma.

"I guess I'm being tested," I replied. "I can't give up and stop doing good things just because bad things are happening to me. If I do, I am the loser. I would be letting the Devil win. Every temptation is a test. Trying to live a good life isn't easy. It may take a long time, and it may not be in this lifetime; but I believe that sooner or later the good you do will come back to you."

CHAPTER 30

That evening I sat thinking about what Lenny had said to me. I knew that I wasn't a bad person. I didn't deserve all the bad things that were bombarding me. Why were they happening?

Then I acknowledged the fact that I had become a thorn in the Devil's side. I definitely wasn't doing anything to place me in his favor. He disliked what I was doing, and he was trying to make my life miserable. I hate to admit it, but he was succeeding. Satan wanted me to turn against God and blame him for my misery. There was no way I could ever do that. But the pain I was suffering was almost intolerable.

I came to the conclusion that if the Devil was really attacking me, I had to do something about it. I couldn't just sit there, do nothing, and let him get away with it. I had to fight back. I had to do my best to stop him.

I decided to call a priest and ask for help. Priests don't receive daily attestation from someone who believes he is being pursued by the Devil, so I was a little hesitant to place the call. I had to tell him everything and I was afraid he would think I was crazy. But I needed spiritual help, and the only way I was going to get it was to place the call and not worry about what the priest thought of me.

When the priest answered the phone, I introduced myself and told him I was in dire need of his help.

"I'm writing a story that is attacking the Devil," I said. "And I think he is attacking me back."

The priest didn't rush me off the phone. He listened patiently as I unloaded all the details of my predicament upon him. To my amazement, he didn't think I was crazy, and he was willing to come to my aid. A blanket of relief descended over me and I began to relax. I was no longer going to be fighting the Devil alone. The priest volunteered to come bless my home and me.

The following day, Lenny worked in the kitchen while I sat sentinel at the living room window anxiously awaiting the priest. Finally he arrived. I dashed to open the front door before he had a chance to ring the bell. When he stepped into the entry hall he was holding a bottle of holy water, a book, and a crucifix. I was relieved to see that he was ready to help me, but I felt a little embarrassed that I had told him such a wild story the day before.

"If I asked you to come here for nothing, I'm sorry," I said apologetically. "But so many evil things have been happening to me I could no longer ignore them. I had to start fighting back."

The priest smiled and looked me straight in my eyes.

"It isn't unusual that when someone attacks Satan, he attacks back. If that didn't happen, there would have never been a need for prayers to stop it."

After hearing his comment, I felt much better.

When we stepped into my living room, the priest handed me the crucifix.

"This is a St. Francis of Assisi crucifix," he said. "It is a gift from me. Hang it somewhere in your house and eventually place a religious article in every room. Evil spirits don't like them."

My living room is similar to a room you would see in an old castle. It has a high dark-wood tongue-and-grooved and beamed cathedral ceiling with heavy dark-wood cross beams supported by corbels below. On the far end of the room is a floor to ceiling fireplace that is crowned with a solid copper hood. The room is furnished, as is the rest of the house, with European antiques.

Like a scene from a Gothic movie, the priest walked from room to room as he recited prayers from the book and sprinkled holy water. I could hardly believe this was taking place in my house.

When the priest first began his ritual, I didn't know what to expect. I wouldn't have been surprised if evil spirits started screaming and furniture started flying around the room. But that didn't happen. There were no sounds or moaning voices and everything remained intact. The priest had come to bless my home and to protect it from evil, not to perform an exorcism. When he finished, I felt that a heavy dark cloud had lifted and his mission had been accomplished.

"Thank you for your help, Father," I exclaimed. "I feel much better."

"You still have to keep up your guard," he replied. "Evil spirits will leave you alone long enough to make you think they're gone for good. Then, when you least expect it, they come back at you in full force. As long as you're trying to expose the Devil, he'll keep his eye on you. You have to finish writing your story. If you don't...the Devil wins."

After the priest left, I went into the kitchen where Lenny was working. I felt much better since the priest had blessed the house, but I sensed that Lenny had mixed emotions about it. He seemed to be a little uneasy. Nevertheless, we proceeded with our work in the kitchen, and I never thought any more about it.

CHAPTER 31

The following day, while I was waiting for Lenny to arrive, the phone rang. It was the mechanic who was working on my car.

"George, you can pick up your car, I finally have it fixed," he said. "This time you can really take it home with you, but I hope you're sitting down because you will never believe what was wrong with it!"

I didn't think I was going to hear anything too surprising. But I soon found out I was wrong.

"What was the problem?" I asked.

"When I first started working on your car, I replaced the plugs and plug wires. Your car kept sporadically cutting out and I've been going crazy trying to find the problem. I checked out everything that could have been causing it a dozen times. I was about to give up when I decided to check the new plugs and wires. One of the new plugs and one of the new wires were faulty. When I went to the parts house to replace them, I discovered the part number for the plug was 666. The part number for the plug wire was 666-0-666."

When I heard that, I almost fell flat on my face. "You're pulling my leg," I said to him. "You've got to be kidding me."

"No," he replied. "I swear I'm telling you the truth."

I found it hard to believe that another incident similar to the "LUCIFUR" license plate had happened. Orland stopped by to visit and I told him what the mechanic said. He too thought it was a joke. We decided to go to the auto supply store and check on the part number ourselves. To our dismay, it was no joke. The mechanic had told me the truth. The part numbers were 666 and 666-0-666. (Notarized documentation signed by Orland DeCiccio is included with other documentation at the back of this book. General Motors can also confirm parts numbers for plugs and wires for a 1981 Buick Riviera.)

What I felt inside is hard to describe. It was a combination of feeling sick, depressed and scared. Then, I thought about the two years of hell I had spent dealing with car problems. Those problems had taken two years of my time away from my book. I wondered if that had been part of the Devil's plan.

I felt as if everything was starting to close in on me, but I knew I had to stay in control. If I fell apart the Devil would win. I had to stay strong and keep my faith in God. I knew I was being attacked by the Devil. But he couldn't win unless I let him.

I must be doing something good or the Devil wouldn't be trying to scare me off and stop me, I thought. *I've got to continue. I can't quit now.*

Orland witnessed the car parts numbers being 666 and 666-000-666. Now, I wanted the search warrant with the name "LUCIFUR" on it. I had thrown the warrant away and I had to get another copy. I called Deputy Richard Hahn, and he got a copy for me. Now I had two things that indicated the Devil had his eye on me. The search warrant with the "LUCIFUR" license plate and the "666" car part.

CHAPTER 32

The day after the priest blessed my home, Lenny never showed up to work in the kitchen. I wasted the whole day waiting for him. Later that evening I called him to see what happened. I can't remember the excuse he gave, but he said he would continue his work on the kitchen the following day. Not so. I wasted another day waiting for him. That evening I called him again. He gave me another excuse and told me he would definitely be there the next day.

"If you aren't going to show up tomorrow, call and let me know," I said to him. "If you aren't coming, I have better things to do than to sit here all day waiting for you to show up."

Lenny reassured me he would definitely work on the kitchen the next day. But that didn't happen either. Now I had wasted three days waiting for him to show up, and I was angry.

That evening I called Lenny again. When he answered the phone he knew I was going to be upset. I barely had a chance to say anything to him before he cut me off. I noticed the resonation of his voice was strange and his attitude was very bizarre. Something was terribly wrong. I felt like I was talking to something evil and I realized nothing was going to be gained by pursuing the discussion any further.

"Just forget it, Lenny," I said. "I'll get the kitchen finished on my own." With that, I hung up the phone and ended the conversation.

About two weeks later, Lenny was still heavily on my mind. I couldn't stop thinking about how strange he had been on the telephone. Plus, I was wondering if he was involved with Carl and Connie. If that was the case, I had to warn him to get away from them. I felt driven to call him once more. Before I could change my mind I placed the call.

Lenny, this is George," I said. "I'm not calling about the kitchen, and I'm not calling to chew you out. I'm not angry."

"Hi, George, what do you want?" he replied in a rather pleasant voice.

"Lenny," I continued, "I'm only calling to tell you something that has been strongly on my mind. If you are involved with Connie or Carl get away from them. I have a feeling they are going to get caught and if you are with them, you are all going down together."

"No, George," he replied, "you don't have to worry about that. I don't have anything to do with drugs anymore and I never see them."

"I hope that's true, Lenny," I responded. "But for some reason those thoughts have been on my mind so strongly I felt I had to warn you. I don't want to see you get into trouble. If you are involved with them, get away from them before it is too late."

Lenny reassured me he had his life under control, and he had no connection with Connie and Carl. Our conversation ended on a good note, but I sensed that he hadn't told me the truth.

CHAPTER 33

Three weeks later, I read an interesting article in the newspaper. There had been a big drug bust in Fontana. A group of people had been busted for manufacturing methamphetamine. My cousin Connie and her boyfriend Carl were both arrested, and Lenny was arrested with them. I felt quite sad inside. If Lenny had listened to my warning, he wouldn't have been in trouble.

Why did I have those thoughts, and why did I feel the need to warn him? I didn't know. But everything I had sensed was true, and everything I warned him about had happened.

I never expected to hear from Lenny again. But two weeks later he called me from the county jail.

"Hello, George, this is Lenny," he said. "I had to call and tell you something that's been bothering me. You tried to warn me. You told me to get away from Carl and Connie or I was going to get caught. I should have listened to you. Everything you said to me was true. But how did you know all that?"

"I'm just as surprised as you are, Lenny," I replied. "I didn't know anything. It was just something I sensed. I couldn't get those thoughts out of my mind and I had to call and warn you."

"There is more you should know," Lenny responded. "It may sound a little creepy, but from the day the priest came to your house I couldn't set foot in your door. You were telling me things I needed to hear, to help me straighten out my life. That priest did get the Devil out of your house, and the Devil made me stay away so I couldn't listen to you. I couldn't go back into your house. That's the reason I never showed up for work anymore. I tried to come back but something wouldn't let me step my foot inside your door."

Lenny was sincere. It had taken a lot of courage for him to make that call. I felt compassion for him and I forgave him for everything he had done to me. His testimony was confirmation that the prayers offered by the priest on my behalf had been answered. I think Lenny learned just as much from that experience as I did.

A few days after that conversation I visited Lenny in jail. He was a much happier person. That spiritual experience had changed his life for the better and he vowed to keep walking on the right path.

Later, I received a call from Lenny's sister, Patty. She told me that her father had disowned Lenny for being involved with drugs. For some reason, I felt compelled to talk with him. I went to Lenny's parents' home, and his father came to the door. After some resistance, I persuaded him to listen to me, and he invited me inside.

I stepped into the house and we went into the kitchen where he poured me a cup of coffee.

"I am finished with Lenny," he said. "He has let his family down for the last time. I don't ever want to see him."

"Lenny needs your support now more than he ever has before," I said to him. "Lenny is at rock bottom. He can't go any lower. The only way he can go is up."

"But Lenny made his own decisions," he said. "It's his own fault he's in trouble, and he has to deal with it himself. There's nothing I can do for him."

"Lenny doesn't feel sorry for himself," I replied. "He knows he did wrong, and he knows he has to pay for it. That's not why he is hurting. He told me that he's suffering the most pain because he knows he has hurt his family. Lenny loves you very much. He needs you, his mother, his family, and his friends to stand behind him and give him moral support. If everyone turns their backs on him, he'll have no incentive to change his life."

I had the poem "The Little Blue Man" with me and I wanted Lenny's father to hear it. He agreed to listen as I recited it. After he heard the poem he told me he really liked it, and by the time I left for home he had a totally different outlook on matters. He loved his son and he was going to be there for him. He changed his mind and said he would talk to Lenny. I hope "The Little Blue Man" and I had something to do with his decision.

A couple of days later I received a phone call from Lenny's sister, Patty, thanking me for what I did. I had helped heal the relationship between her father and her brother. I felt very good about that.

With God's help, my cousin Connie and Lenny turned their backs on drugs. They freed themselves from the grip of Satan and are now good hard working productive people. God is back in their lives. Connie's brother Doug, who went to prison for manufacturing drugs, also got his life in order with God. He is now working to help other people see the light.

CHAPTER 34

For the balance of 1992, I continued working on the screenplay. The Devil continued working on me. I was resolved to not crumble under any circumstance, and I finally managed to complete the first draft of the script. Now I was ready to start on the rewrite.

Chris Costello, the daughter of Lou Costello, told me her nephew was a writer and he had successfully written several things that were very good. He also knew how to correctly write a screenplay. His name is Michael Cristillo, "Cristillo" being the real "Costello" family surname. I soon met with Michael and hired him to work with me on the script.

Amy, the main character in my story, has a music box that was given to her by her father just before he died in a car accident. The music box is a major part of the story, and it plays in several scenes throughout the script. It was always in the back of my mind that the music box should play its own original tune.

One evening I was driving home from an event I had attended with Jane Withers. As I was buzzing along the freeway I started to think about the music box. Suddenly I heard a melody playing in my mind. Part of it was from a song I had written years earlier, and part of it was new. I made changes in the tune and played it over and over in my mind all the way home. By the time I pulled into my driveway I had the tune I wanted for the music box. I immediately went into the house and picked it out on the piano so I wouldn't forget it. I fine-tuned the song and then I called my friend Betty Rose.

Before I go any further there are a few things I must say about Betty. Our mutual friend, Jane Withers, introduced Betty, her husband David, and me to one another in 1974. I have had great admiration for Betty from the moment we met. She exudes a sincere warmth and goodness, and she lives up to the image she projects. Of all the people I have met in Hollywood, Betty is one of my favorites. I am happy to say my opinion of her is not unique. She is loved by all who know her.

Betty's husband, David Rose, was one of Hollywood's top composers. He composed the musical scores for many motion pictures, and he also did the soundtracks for "Bonanza," "Little House on the Prairie" and "Highway to Heaven." One of his biggest hits, "The Stripper," is, and will forever be, the staple song of any striptease act. David passed away several years ago, but his music will live on forever.

When I called Betty, I thought perhaps she would know of someone who could write my song down on paper. I had the tune for the music box, but I didn't know how to write music.

"Betty," I said, "this is George. I've written a song for the music box. I need sheet music for it but I can't read or write music. Do you know anyone who can help me?"

"Of course," Betty replied without hesitation. "My friend Pete Rugolo can do it. I'll call him, and we can all get together here at my house."

Pete Rugolo had his own orchestra and had created the musical scores for the TV series, "The Fugitive." He had also scored other television shows as well as many motion pictures. Pete's accomplishments were endless. Never in my wildest dreams did I think someone of his caliber would take the time to work on my song. When Betty called to tell me Pete had agreed to do it, I was thrilled. For a change, something good had happened to me.

Within a few days, I met Pete at Betty's house. I played the song on the piano and he wrote the music down on paper. A couple of weeks later we met again at Betty's home. Pete had written an arrangement on my song and the sheet music was finished. He even had a friend of his make a demo tape of the music. When I heard it I was so pleased tears came to my eyes.

During the next few weeks, I wrote lyrics to the song and Pete revised the arrangement and rewrote the music to accommodate the words. The title of the song is "Those Were the Times to Remember." I had a demo tape made with vocal, instrumental, and music box versions of the song. When Pete and Betty heard the tape, they told me they thought it could become a hit song. I hope they were right.

Words can't express how much I appreciate the help Pete Rugolo gave me. He could have charged me thousands of dollars for his work, but he refused to take one cent.

"I did it as a favor," Pete said. "I don't want any money for it."

I will always be thankful to Pete and Betty for helping me and for being so generous and kind.

CHAPTER 35

Plenty of bad things were in the works elsewhere. Problems with tenants and monetary losses at the apartments both increased. Every tenant ripped me off. Not one of them kept their word or paid their rent.

I remember one tenant especially. Her name was Ginger. I happened to be at the apartments when she stopped by to see if I would rent to her. She was well dressed and appeared to be very stable. She told me where she worked, gave me references and basically pulled a first class con job on me.

I didn't have a rental agreement with me and she didn't have her deposit with her. She asked me if I would take $100 and hold the apartment for her until she went home to get the rest of the money while I went to get a rental agreement. I wanted to think she was being honest with me, so I agreed. She then asked me if I would give her the key to the apartment so she wouldn't have to wait outside when she returned. I didn't think that would be a problem, so I handed her the key. She left to get the rest of the deposit money and I left to get the rental agreement.

When I returned with the rental agreement, Ginger wasn't there. I waited for several hours and she didn't show up. I had other things to do, so I left. She made no attempt to get in touch with me.

After several days, I stopped by the apartments to find she had moved in. I asked her for the money for the balance of her deposit and her first month's rent. She apologized for not having the deposit or the rent money with her and told me to come back. Needless to say, for the next month I didn't hear from her. Every time I went to the apartments to see her she was never home. I told the tenants in the front apartment what was going on, and they agreed to call me as soon as Ginger returned.

About a week later, I received a call from the tenants and I drove to the apartments. When I confronted Ginger, she gave me another excuse. I don't remember what it was, but she begged me to give her until the end of the month. She promised me she would then pay the rest of the deposit and all the rent. Again I waited. The end of the month came and I still had no contact from Ginger. She didn't pay her deposit and she didn't pay her rent.

It was two weeks later before I was able to find her at home. When I confronted her about the deposit and rent money she started to give me another excuse. But this time I was angry. I told her she had to pay me within three days or I was going to evict her. Three days passed. I received no deposit and no rent.

I mailed Ginger a legal notice to pay or quit. She still didn't respond, so I went to court. She appeared at the hearing and stated that something was wrong in the apartment and I wouldn't fix it, therefore she had reason to not pay the rent. That lie is what most deadbeat residential tenants use to delay an eviction. But this time it didn't work. Ginger was ordered to vacate the apartment.

A few days later, I received a phone call from another tenant. She told me Ginger was stealing anything she could get out of my apartment. She had even confronted Ginger and asked her what she was doing.

"This is nothing," Ginger replied. "I'm just getting started. Wait till I'm finished. I'll fix his ass good."

By the time I got to the apartments, Ginger had moved out. She had stolen the ceiling fan, light fixtures, and even the security screen door. She had torn up the kitchen cabinets, smashed walls, destroyed the alarm system, and wrecked the inside of the apartment.

I called the police and filed a police report. Ginger's neighbors told the policeman they would testify that they saw her carting the stolen items from the apartment. And, they would affirm the comments Ginger had made to them. Finally, someone who ripped me off wasn't going to get away with it. I told the policeman I wanted to press charges against Ginger.

A few weeks later, I received a letter from the District Attorney's Office. I was informed they didn't have enough evidence to prosecute. I was shocked. How could that be possible? There were witnesses willing to testify against Ginger. How could the District Attorney refuse to prosecute because of lack of evidence? I couldn't believe it.

I couldn't believe the legal system was not going to help me. To prosecute Ginger, I would have had to hire an attorney and fight it out in civil court. I couldn't afford attorney's fees and court costs so I had to accept the fact that evil had won. I didn't have the money to buy justice. Needless to say, I became disillusioned with our legal system.

The loss of rental income, repairs to the apartment and the replacement of stolen items cost me thousands of dollars. Ginger lost nothing. The legal system allowed her to gain from her evil. I was the loser.

The system didn't fight for me. But if anyone had done the same things to a government building or agency that Ginger did to me, they would have been arrested and prosecuted.

CHAPTER 36

The incident with Ginger is only one of many I dealt with. Owning that triplex had become a living nightmare. I was constantly dealing with one problem or another. I was also running out of money to cover the losses. I had a second mortgage on the house, a second on the apartments and my credit cards were almost charged to the maximum. I lost over one hundred thousand dollars.

I was starting to feel as if everything was closing in on me. I couldn't handle any more stress and I was suffering from unbearable depression. I couldn't sit still and I couldn't sleep. I felt like a kernel of popcorn on a hot griddle, bouncing around getting ready to explode. That's when it all came to a climax.

I received a phone call from one of my tenants. I was informed that another tenant had moved out and burned me for back rent. That wasn't all. They had also demolished their apartment. My mind couldn't handle more problems. I felt as if my brains were exploding inside my head. I threw down the telephone and I started screaming hysterically.

"God, why are you doing this to me? I don't deserve this! I've had faith in you! I'm writing this story to expose the Devil and you are letting this happen to me. Damn you! I don't deserve it!"

I pulled my hair and paced through the house ranting and raving. I lost all control of my senses. Randy was in his upstairs room when he heard all the commotion. He missed half the steps as he dashed down the stairway to see what the ruckus was all about. By the time he got to me I had totally lost it.

"George! Calm down," he yelled. "Stop screaming and calm down!" He grabbed me and tried to shake some sense into my head, but I kept screaming and shoved him away.

"Leave me alone!" I screamed. "The seed you sow is the harvest you reap, and, what goes around comes around, are nothing but damn lies. God is nothing but a damn liar." I looked up toward the ceiling and continued to scream at God. "You damned Son-of-a-Bitch in the sky. You are nothing but a damn liar," I yelled. "I don't deserve this! Why are you torturing me?"

"George, don't talk like that," Randy decreed. "You can't mean what you said."

I leaned over the sofa with my face in my hands and sobbed until I couldn't cry anymore. Randy tried to calm me down, but he was going out for the evening and was only there a short while longer.

By the time I finally regained my sanity, I was alone. I walked around the house feeling deep shame for the terrible things that I had said to God. I don't think I have ever felt so low in my life. I had screamed at God and called him names. Now that I had regained my senses I felt like the scum of the earth. I was extremely sorry, but it was too late.

I walked around the house for a few more minutes. I felt worse with every step I took. I walked into the living room and looked at the St. Francis of Assisi crucifix the priest had given me. It was hanging on the hood of the fireplace. I fell to my knees in front of the crucifix and started to pray.

"Dear God in heaven, please forgive me for the terrible things I said to you. I love you, God. In anger I lost control and I screamed things at you that I didn't mean. I'm not worthy of your forgiveness but if it's your will, please forgive me. I feel terrible for what I said, God. I am so sorry. I've tried to be strong but I've had nothing but bad things happen to me. Please forgive me. And please, I beg you to give me a sign so I know what's going on and why all these things are happening. I'm at rock bottom and I can't handle any more."

The telephone rang. I stopped praying and got up to answer it. It was my neighbor, Mary Ellingwood.

"What are you doing," she inquired. "Have you had dinner yet?"

"No," I replied.

"Why don't you go with me to get some Chinese food," she asked. "We can bring it back to my house and eat here."

I knew it would be better for me if I weren't alone that evening so I accepted Mary's invitation.

"Sounds good to me," I replied. "I'll be right over."

Mary and I drove to the Panda Inn, picked up the Chinese food and drove back to her house. She made tea and we sat in her living room while we ate our dinner.

Mary told me all about her day. She was having financial problems of her own but she was dealing with them. By the time we finished the main course, she had filled me in on all the details of her dilemma.

"So, how was your day," she asked.

Mary, as well as most of my neighbors and friends, knew about all the strange things that were happening to me. I told her the rotten truth of what I had done just prior to her phone call to me. I told her how I had gone berserk and shouted all those nasty things to God.

"After I calmed down I felt terrible for the awful things I screamed at God," I said. "I said a prayer and asked for his forgiveness. Then I asked him if he would give me some sign so I would know why all these bad things were happening. I told him I was at rock bottom and couldn't handle any more."

As I finished that sentence, Mary handed me my fortune cookie. I removed the wrapper from the cookie, cracked it in half and pulled out the paper fortune slip that was sealed inside. When I read what it said, I was so astonished that I had to read it again. Mary could tell from the expression on my face that I was flabbergasted.

"What does it say," she asked.

"Maybe this is the sign that I prayed for," I replied. Then I read the message on the slip of paper. "Among the lucky, you are the chosen one."

I couldn't believe my eyes and Mary couldn't believe what she had heard.

"Come on, you're kidding me," she said. "You made that up."

"No. That's really what it says," I replied.

I handed her the fortune slip and she read it herself.

"George, I'm out of here," she kidded. "Too many strange things happen to you."

"Maybe this is the sign that I asked for," I responded. "That message wouldn't mean anything to anyone else."

When I returned home that evening I couldn't get the fortune slip off my mind. I felt that message was meant for me. I thought perhaps I had been chosen to write my story, and perhaps that was my destiny. I was exposing and condemning the Devil, and he was angry. He didn't want my story to be told, so he had to destroy me. He was the cause of my anguish and he had tricked me into blaming

God. I had been weak, I had fallen into the Devil's trap, and I had done exactly what he wanted me to do.

When I was yelling at God, the Devil must have been gloating over his accomplishment. At that moment I had let the Devil win.

The message on that fortune slip made me realize that writing my story might be my God-given destiny. And, as long as I was attempting to fulfill that destiny, the Devil was going to make my life miserable. From that moment, I have never blamed God for my misfortunes.

I placed the fortune slip in my wallet so I would always have it with me. I showed it to everyone who knew about my story and I told them the events that occurred the evening I received it.

From the moment I was handed that little piece of paper, I have felt at peace. But my problems didn't stop. I only had the fortune slip in my wallet for a few months before my wallet was stolen.

I was more upset about losing the fortune slip than I was about losing the wallet or anything else that was in it. Nothing I could do was ever going to get that fortune slip back. I had to deal with its loss and get over it.

That fortune slip was only a piece of paper but it gave me gifts that can never be taken from me. It intensified my faith in God and empowered my conviction to fulfill my mission. To me, it was the answer to my prayer.

CHAPTER 37

Michael and I continued to work on the rewrite of the screenplay. The second draft was completed in March of 1993.

A few weeks later, on the night of the Rodney King riots, I was driving home from the Chino store. It was about nine o'clock in the evening. Oncoming headlights hitting a film on the windshield of my car were creating a blinding glare. I was near a do-it-yourself car wash so I decided to stop and clean the windshield. I pulled into one of the wash stalls and inserted money into the slot. Then I decided I might as well wash the entire car.

As I sprayed the car, I noticed an old red car move slowly through the drive behind the car wash. I rinsed the car and began wiping the water off with a towel. Then I noticed a black boy sitting behind the car wash on a dimly lit vacuum tank. I continued wiping the car. The next time I looked behind the car wash, the black boy was gone. Within seconds I heard a noise in the stall next to mine. It sounded like someone had tripped on something. There was no car in the next stall and I immediately felt something was wrong.

I was wearing a lot of jewelry. I had a three-carat diamond solitaire on my left hand, a nugget bracelet with fourteen quarter-carat diamonds on my left wrist, a large gold nugget with two diamonds totaling thirty-five points hanging on a gold box chain around my neck and a gold ring on my right hand.

I knew I should take the jewelry off and hide it. I quickly took the ring with the large diamond off my finger and put it in my shirt pocket. I started to remove the bracelet but it was too late. Two black boys charged into my stall so fast that they startled me. My heart skipped a beat, but for a moment I wasn't afraid. They stood near the back of my car for a few seconds and watched me. I looked at them and smiled.

One boy was short, light skinned, and had a round face. He appeared to be around fifteen years old. The other one appeared to be around eighteen to twenty years old. He was tall and skinny. He had straggly hair and a little goatee. He looked like a dirty, undernourished rat.

"Gee, you guys scared me," I said.

I was uneasy when neither of them responded. The tall, ratty looking character raised his hand to make sure that I saw he was holding a gun. My heart felt like it missed ten beats. I didn't have to say a word. My body language told him that I had seen it.

"Get up against the wall," he demanded.

Within seconds I was against the wall. His comrade in crime asked for my wallet, and naturally, he got it. There was only ten dollars in it. He took the ten dollars and, to my surprise, he handed the wallet back to me.

"Give me your necklace," he demanded.

When I raised my arms to remove the necklace, the bracelet that had been hidden by the sleeves on my shirt was exposed. He saw it; he wanted it, and he got it too.

I could hear the voices of people standing outside a fast food restaurant next door to the car wash. The grungy gunman stood on the outer edge of the stall and kept one of his evil eyes on the people and the other eye on me.

"Don't you dare move," he demanded, as he pointed his gun at me. I wanted to do as he asked, but I couldn't. I was terrified and I was shaking uncontrollably.

Then the short round-faced gangbanger ordered me to lie on the floor and put my head under the car. That is when I really got frightened.

They're going to shoot me, I thought, *and I'm not even going to have a chance to say goodbye to my kids.* I prayed silently, "God, please help me. Please don't let them shoot me."

I tried to keep my composure. I didn't want do anything that would antagonize them and provoke them to shoot me.

I can't believe this is how my life is going to end, I thought. *If I die now every unexplainable incident I've experienced would be without purpose. I can't believe this is how my life is going to end.*

The floor of the car wash was covered with water puddles, but getting wet was the least of my worries. I got down on my hands and knees and started to lie down on the floor. To my surprise, both thugs ran out of the car wash and vanished into the night. I felt a tinge of relief, but I was still intensely afraid.

The second I knew they were gone, I shot up from the floor and jumped into my car. I slammed the door and started fumbling with the keys. My hands were trembling so hard I could barely hold on to them. I felt like I was going to have a stroke.

I finally turned the ignition key and the car started. I was afraid those thugs might be watching me, but I didn't have time to worry about it. All I cared about was getting away from there as fast as I could. I threw the car into gear and slammed my foot down on the accelerator. I thank God no one was on the sidewalk and no cars were nearby when my car careened into the street. I was so scared, I didn't take time to look. I was a half block down the street before I started to think with reason.

I drove to a gas station on the corner. I was still so frightened I couldn't think rationally. I didn't know if the thugs were gone or not. I jumped out of the car and darted inside the station.

"Call the police!" I yelled at the clerk. "I've just been robbed by two armed gangbangers!" He placed the call and within minutes a squad car pulled into the drive of the station.

When the officer got out of his car I could barely explain what had happened. The thieves had taken over $10,000 worth of my jewelry, which was uninsured, but they missed two valuable things. The ring in my pocket was worth three times that amount. They also missed a $2,380 cash bank deposit in the car's glove box. But if they had taken that too, I wouldn't have cared. At that moment money and jewelry meant nothing to me. My main concern was that I was alive. I could always replace the material things.

That entire night I remained in a state of shock. I paced the floor from one room to another and relived that experience a hundred times. The horror I felt when that gun was pointed at me ricocheted throughout my body.

Why did this happen? I kept thinking. *I've finished the script. There shouldn't be any more bad things happening. They should have stopped.*

When the sun came up the next morning I was still pacing the floors. But I had accomplished something. I had calmed down enough to think more clearly. I realized why that robbery had taken place. It was meant to happen.

Every strange event I had ever encountered had its purpose. I was living a part of my own story. The near-death experience, the dream that started the story, the "LUCIFUR" license plate, the "666" car part, and every other encounter—supernatural or otherwise—was meant to happen.

Amy, the main character in my screenplay, was basically doing everything I was doing in my own life. In my story she was writing a screenplay. In real life, I was writing a screenplay. Amy's mission was the same as mine; exposing the Devil, warning people to get their lives in order with God, and trying to stop some of the violence in the world. Without realizing it, I had been writing a fictitious script, but living the true story the script should be based on. The only things missing from my script were the unbelievable events I had confronted.

I knew I had to do another rewrite. My near-death experience, the dream that started my mission and all the strange things that I had encountered belonged in my script. Those things had to happen to Amy. My story had to be based on my own life.

CHAPTER 38

I called Michael Cristillo and told him what had happened. Soon, the third rewrite began. At last I was going to be pleased with my story. It was going to be based on supernatural, mysterious, and unexplainable true events from my own life.

The search warrant had to be included in the story, so I got it out of the drawer and started to read it. For some strange reason a thought popped into my mind. *I wonder if there is a message to this 1972 Ford and this "Lucifur" license plate that I am missing?*

I had never done anything with numerology, but I knew how it worked. I decided to convert all the letters in "Ford" and "LUCIFUR" to numbers. When I started adding and subtracting different combinations of numbers, the answer to every combination was 1998. Six different combinations, and they all totaled 1998.

What does 1998 mean, I thought. Then, it was as if a voice said to me, "1998 will be the third time something will occur since Christ died." I then divided 3 into 1998. The answer was "666." 1998 is the third multiple of 666. I later discovered that 1998 divided by 6 is 333, the trinity. Then I decided to add "1972" plus the numbers allotted to the letters for Ford, "6694." That total was "8666."

I couldn't believe what I had discovered. The search warrant was also on file at the county courthouse and at the Sheriff's Department. There was proof I had never owned the 1972 Ford with the "LUCIFUR" license plate. That license plate had been registered to that car for nineteen years. No one could have possibly known that car and plate were eventually going to be part of a numerical configuration that would equal "666." It could not have been pre-planned. If "LUCIFUR" had been spelled right, it would not have worked. If Judge Ellen Brodie had written the right date, it would not have worked. Any year other than 1972 would not have worked, and any car other than a Ford would not have worked. I also used the year "1982," the year that the ownership of that Ford was transferred to someone in Compton. Any year other than 1982 would not have worked. Everything had to be just as it was for the totals to equal 1998 and 666. (A copy of the numerological chart is included with other documentation at the back of the book.)

I immediately called my friend Orland. When he arrived I showed him what I had discovered. He was shocked. We both knew that something far above any power on earth was at work. This was no accident. I didn't fully understand what was happening, but I realized I was involved with something very serious. This was definitely no game.

"What do you think all of that means," Orland asked.

"I don't know," I replied. "But I think a major occurrence will take place in 1998, and it will involve the Devil or some form of evil."

About a week after my "1998" and "666" discovery, I met with Michael Cristillo and showed him the numerological chart I figured out. I'm sure Michael had a hard time believing some of my stories. But, we continued our work on the script and soon, with even more changes, the third draft of the script was completed.

CHAPTER 39

Things seemed to calm down. Other than problems with the apartments, my life was going fairly well. I was even surprised by a tidbit of news from my friend Orland. He had told a private group of investors about my screenplay. They were interested and had requested a copy of the script. Naturally, I couldn't get a copy to them fast enough.

A few weeks later, they contacted Orland. They liked the script and they were discussing backing a production of the film. I was happy other people liked my story, but I was thrilled that there was a possibility of getting funding for production. I anxiously awaited their decision.

During that waiting period, I was working at my Rancho Cucamonga store. One day I received an unexpected call from my friend Brenda Wallis. Brenda was the friend I had met working in the restaurant next door to the barbershop I had worked in during the early 1970's. She had moved to the Sacramento area, but she was in Riverside visiting her son and daughter-in-law.

"George," Brenda said. "I'm at Dean and Sandy's home in Riverside. They're having major financial problems and they're emotionally and financially devastated. They've stopped going to church and they're blaming God for everything that's happened to them. If I bring them to you, will you tell them what you've been through? I think it will do them good to hear your story. It may change their thinking."

"Bring them to the store." I replied. "I'll be happy to tell them everything I've experienced. I think I can help them see where their troubles are really coming from."

Brenda, Dean and Sandy arrived at the store about an hour later. The last time I had seen Dean he was a young boy. Now he was a grown man and had been married for several years. He and his wife were both around twenty-six years old.

Between waiting on customers, I told Dean and Sandy the story of my life. They listened intently as I spoke about the near-death experience, the dream that started my mission, and my ongoing battle with the forces of evil. I hoped by telling them the bad things, the good things, the supernatural things, and the lessons I learned, they would be enlightened.

"I made a big mistake," I said. "Many bad things were happening to me. Like you, I was devastated. Like you, I started to blame God. I was doing exactly what the Devil wanted me to do. The Devil constantly torments us because he wants us to blame God. If we do blame God, the Devil wins."

Finally, I told them about the night I screamed at God and blamed him for my problems.

"I felt horribly guilty for the things I said to God," I told them. "I prayed and asked God to forgive me, then I asked him to give me a sign so I would know why all those things were happening to me. I was at rock bottom and I couldn't handle anymore."

Then to wrap up my story, I told them about the fortune cookie. And, the message that said, "Among the lucky you are the chosen one."

The second Sandy heard about the fortune cookie she raised from her seat.

"Oh my God! I can't believe this," she exclaimed.

I didn't expect to get that kind of reaction. For a moment I thought she didn't believe me.

"What do you mean?" I asked.

"When I was seventeen I got a message in a fortune cookie that said the same thing," she replied. "I kept it, and for years I've looked for another one. I never found one until last week. I have it in my purse, and I think it was meant for me to give it to you."

Sandy reached in her purse, pulled out the fortune slip and handed it to me. I looked at the slip and read it aloud.

"Among the lucky, you are the chosen one."

"Oh my God, this is unreal!" Brenda remarked. "I have goose bumps all over me."

Brenda wasn't alone. Dean, Sandy and I had chills. We all knew that, miracle or not, we had been participants in an extraordinary occurrence. I was stunned that I had received that message on a fortune cookie slip for the second time.

CHAPTER 40

The events that took place that day were an inspiration to all of us. Dean and Sandy realized where their problems were really coming from. They stopped blaming God and started blaming the Devil. Brenda was happy that their faith in God had been renewed, and I was elated that I had helped them get back on the right path.

I was thrilled that I was given another fortune cookie slip to replace the one that had been stolen. The new one means even more to me. I keep it in a safe place. Whenever I get discouraged or depressed I take it out of its container and I read the message over and over. Those words always regenerate my faith in God and my confidence in myself.

A few weeks later, Jane Withers called and invited me to dinner. She wanted me to meet a friend of hers who was the hostess of a talk show in Raleigh, North Carolina. I met Jane and her friend, Dea Martin, at the Portofino Restaurant in Sherman Oaks. After dinner, Jane wanted me to tell Dea everything that had happened to me, which I did. I also recited the poem I had written about "The Little Blue Man."

Dea loved the poem and was so interested in my story that she asked me if I would fly to Raleigh and be a guest on her television show. I was ecstatic that someone I had never met before had such an interest in my story and also loved the poem. Of course, I wanted to be a guest on Dea's show, so I agreed to make the trip to Raleigh. The show date was set for September 2, 1993.

On August 14th, 1993, Morton Downey Jr. was honored at a ceremony at Gower Gulch. Gower Gulch is a western-theme shopping complex located at the corner of Gower and Sunset Boulevard in Hollywood. I attended the event with Venice "Richie" Wildberger, president of the Film Welfare League; a Hollywood-based entertainment industry charity. About 20 guests were there to watch Morton place his autograph and handprints into the cement sidewalk.

Everyone was chattering and waiting for the ceremony to begin. I was standing next to an attractive well-dressed blonde lady. One of us managed to say "hello" and we were soon involved in full-blown conversation.

"What do you do?" she asked.

"I own and operate two formalwear stores," I replied. "But I have also written a screenplay. I have potential investors interested in it, and I am trying to get it produced."

"Really," she said inquisitively. "What is it about?"

"Well," I replied with a little hesitation. "It's based on a bizarre true story. Its main message is that Satan is controlling the world, his time is short and he is trying to get every soul he can get before his time is up. That is why evil, drugs and corruption are spreading around the world. The screenplay is about a girl who is trying to warn people to get their lives in order with God before it's too late. She is working for God but she has angered the Devil and he constantly pursues her. His goal is to destroy her so her story can never be told."

I could tell from the lady's inquisitive expressions that she was very interested in what I had just said.

"You said the screenplay is based on a true story?"

"Yes," I answered. "Most people will find it hard to believe but it is based on true things that happened to me."

"What kind of things happened?" She asked.

I paused for a moment to think of how I could tell her the basic facts and not sound crazy. At one time I would have been more concerned about other people's opinions, but I had learned and grown from my experiences. No matter what people would think, I had to tell the truth.

"I had a near-death experience and a dream that started my story," I replied, "and ever since I started writing it, I have been under attack by the Devil. There are a lot of supernatural things that happen to the girl in the script, and most of them happened to me."

I told her the details of the near-death experience, the dream, the "LUCIFUR" license plate, and the number "666" on the car part.

"How do you handle being under attack by the Devil?" she asked.

"I didn't know what to think at first," I replied. "I was a nervous wreck. But since I realized the reason it was happening, I haven't let it bother me. I have to stay strong."

"But aren't you afraid of the Devil?"

"No," I replied, "but the Devil is powerful. One of his biggest tricks is convincing people that he doesn't exist. Another one is getting people who believe he exists to be afraid of him. If they fear him, he can manipulate them to do his will. All he has to do is create havoc in their lives and, because they fear the Devil is after them, they stop the good they were doing. They lose and the Devil wins. He can do anything he wants to me, but he can't get to me unless I let him. As long as I'm not afraid of him, and my faith in God doesn't weaken, he can't win."

"But how do you handle all the bad things that happen to you," she remarked. "Don't you wonder where God is and why he allows bad things to happen?"

"Our faith in God is tested every day," I answered. "If our faith weakens the least bit, the Devil knows we're vulnerable and he'll never leave us alone."

"But why doesn't God stop the Devil?" she asked.

"We are tested by God too," I responded. "God wants us to have faith in Him through thick or thin. If we have faith in Him only when things are good, He knows our faith in Him is weak and we're easy prey for the Devil. If we retain our faith in Him when things are bad, He knows our faith is strong and we're in His favor. The Devil may succeed at making us miserable, but he doesn't have a grip on our souls. When our intentions are to do good the Devil will always try to stop us. We have to stay strong and fight him. We can't let him win."

"This is totally unbelievable," she responded. "I didn't really want to come here today but for some reason I was motivated to get ready. I think I was supposed to be here so I could hear what you just said. You'll never know how much I needed to hear it."

"Really?" I queried. "I thought you might think I was some kind of nut and walk away. But, I swear everything I said to you is the truth."

"You don't have to convince me," she responded. "I know the Devil exists. I think he's been attacking my family and me. Many bad things have been happening and I can't handle anymore. But you're right. I have to stay strong. I can't let him win."

By this time my curiosity was going wild. I could hardly believe someone else could relate to the things I had been dealing with for so long.

"By the way," I said, as I extended my arm to shake hands with her. "My name is George Newberry."

"I'm Karen Kramer," she replied, as we shook hands. "My husband is Stanley Kramer."

I instantly recognized her husband's name. Stanley Kramer was the producer-director of many major motion pictures. "Guess Who's Coming to Dinner" and "It's A Mad Mad World" were two of his many successes.

I was overwhelmed to learn I was talking to Stanley Kramer's wife, and amazed that she was so responsive to my story. We had so many things to share with one another that we couldn't talk fast enough. We were so involved in our own discussion that we didn't even notice the ceremony had started, until someone asked us to be quiet and listen. Needless to say, that put an abrupt end to our talk.

After the ceremony, Karen told me she would like to meet with me at some future time to continue our discussion.

"I also want my daughter Katherine to hear this story," she said. "She needs to hear it just as much as I do."

We exchanged phone numbers so we could call and set a date, time, and place for our meeting.

"I will bring a copy of my script if you would like to read it," I commented.

"Call me soon, I would love to read it," she replied.

I was exhilarated. I had no idea I was going to meet someone at that ceremony that would be interested in my experiences, and maybe even interested in my script.

CHAPTER 41

The following week I flew to Raleigh to appear on Dea Martin's television show. Michael Cristillo and I had completed the third draft of the script, so I arrived in Raleigh with a copy in hand.

The day of the taping, I was a nervous wreck. I had been on television several times as a guest at televised functions in Hollywood, but I had never had to speak on camera. To make matters worse, the show was being broadcast live. I knew if I goofed up on camera there was no way to fix it. There would be no second chances.

"God, help me," I prayed. "I don't want to make a fool of myself. Please help me do well on camera."

We had no script and had little rehearsal for the show. Dea introduced me, started asking questions, and I answered. I told the viewers my story and I wrote the numerological configurations that I had figured from the search warrant on a black board. To my surprise, I didn't make any mistakes on camera, and I wasn't that nervous.

Following the show, I was the guest of honor at a luncheon held at the Raleigh Holiday Inn. I was surprised to see so many people attending. They had all watched the show and were interested in my story. I enjoyed talking with them very much, but I did feel strange when people asked for my autograph. I was not a celebrity. The audience response was so great I was asked to appear on the show again the following day.

On the second show I played a tape of the song I had written for the music box, "Those Were the Times to Remember." The song is a real tearjerker. Many people relate to the lyrics and are very touched by them. I'll never forget how much Theresa, an employee at the television station, cried when she heard it. I was deeply moved by the comments of the studio staff and was elated that the song was a major hit to them.

Dea was delighted with the audience responses to both telecasts. They were two of the most popular telecasts during the entire run of her show. It also proved to be a good experience for me. With tapes of the shows in hand, I left North Carolina with more confidence in my story, my song, and myself.

CHAPTER 42

As soon as I returned to California, I called Karen Kramer. I had been looking forward to our meeting, and she was anxious to talk with me.

A few days later I drove to her home in Studio City. Karen and I talked for quite some time. She told me some of the bad things that had happened to her family. One thing I remember was that a man had ripped them off for millions of dollars and they had to sell the mansion they had previously lived in. She also told me they heard noises and someone whistling in the house, but no one was ever there.

I again told Karen how I had dealt with my own problems. I did my best to convince her to ignore all the bad things that were happening and to keep her faith in God.

Eventually Karen felt better about her situation, and I showed her the videotape of the two shows. She loved my song and she was mesmerized by my story. Katherine was out for the afternoon and Karen asked me if I would wait until she returned home so she could hear my story. Of course, I waited.

When Katherine returned, I repeated my story. She was just as responsive as her mother had been. They both went on and on about how thrilled they were to know me. I felt as if I had known them for a long time. I spent most of that afternoon and evening with them talking about the story. Before I left for home I handed Karen a copy of my script.

"I can't wait to read it," she said. "I'll call you as soon as I'm finished with it."

About two weeks later I heard from Karen. She had read the script and she loved it.

"This story is wonderful," she said. "This movie has to be made. Can you bring me four copies of the script and four tapes of the song? There are several people I want to talk to about it."

I immediately delivered the scripts and tapes to Karen.

"I would like to see Katherine play the part of Amy," she said. "She is perfect for that role."

I had never seen Katherine act, so I had no opinion about her capabilities. But I figured if Karen was able to get a producer to make the film, Katherine could play the role of Amy.

"I have an appointment with Irwin Allen," Karen said. "I want him to read the script."

I was excited that Karen liked the script well enough to take it to Irwin Allen. I knew who he was; he had produced some very good films and television shows.

I didn't know if he was even going to like the script, but all the way home I fantasized about getting the movie made.

For several months after that meeting I had no communication from Karen, but I received several calls from Katherine. She seemed to think I was a psychic or a psychologist. She was obsessed with Faye Dunaway. She desperately wanted to meet her and establish a friendship.

Almost every day I received a call from Katherine asking me what I thought would be the best time for her to contact Faye, or if I even felt it was the right day for her to make the call. I finally told her I was not a psychic and I had no way of knowing when or what she should do.

CHAPTER 43

Weeks passed, and Karen never called to give me an update on the developments with the script. I knew things took time and I didn't want to be pushy but my curiosity got the best of me. I had to know if she had received any communication from Irwin Allen.

I called Karen several times and left messages on her recorder. A couple of times her housekeeper answered the phone. I asked her if she would please tell Karen to return my call. I still didn't receive a response.

I thought it was strange that Karen didn't return my calls but I had no reason to think anything was wrong. I assumed she must have been busy and she hadn't heard anything from Irwin Allen.

About a week later, Orland received some news from the investors. They told him they were still interested in the script and would notify us of their decision soon.

I again called Karen. Again there was no answer. I left a message on her recorder telling her that I had received some news from the potential investors. This time Karen returned my call.

When I answered the telephone, Orland happened to be at my house. Karen informed me that the reason she hadn't called previously was because she had been too busy. However, she was glad to hear that we had heard from the investors.

Since Orland was at my home I had Karen talk to him. He told her what the investors had to say. Karen told both of us she intended to see that the film was produced, and that Irwin Allen was interested in making the movie.

After that call, the lines of communication were again open between Karen and me. She told me she had been going through lots of bad experiences, and she felt like the Devil had been doing double duty on her.

CHAPTER 44

On January 17, 1994, at 4:31 a.m., the Northridge earthquake struck Southern California. I was awakened by the sound of the house popping and cracking. In an instant, I was out of bed. The house was rocking back and forth like a boat on water. I ran into the entry hall to seek safety. I held tightly onto the doorjamb and prayed the quake would subside.

I thought the quake would stop any second, but it didn't. The shock waves increased. I was scared. Walls cracked and furnishings started crashing onto the floor. I kept praying that God would stop it. A few seconds later, it was over.

I held onto the doorjamb and waited a few more seconds to see if it was going to start over again. When I was fairly sure the worst was over, I started walking around the house and looked for damage. I was lucky. Most of the things that had fallen onto the floor hadn't broken. However, there were cracks in the ceilings and walls that needed to be repaired. I was thankful the quake hadn't lasted longer. Even a few more seconds of shaking would have caused major damage.

I knew everyone in Southern California had to feel the quake. I turned on the TV to watch the news reports. Major damage had taken place in the San Fernando Valley. Buildings had collapsed and people had been killed. I had been fortunate. A few cracks in my house meant nothing compared to the trauma other victims of the quake had to face.

The thought kept coming to mind that somewhere in the Bible it says in the end time there will be earthquakes that are warnings.

"Was the Northridge quake a warning," I kept asking myself.

I constantly thought about it. Then one day I decided to do numerology on the month, day, and time of the quake. January 17th, at 4:31 a.m.

I took January, and converted each letter to its number in the alphabet, adding double-digit numbers together. J the 10th letter, 1 + 0 = 1. "a" the 1st letter = 1. "n" the 14th letter, 1 + 4 = 5. "u" the 21st letter, 2 + 1 = 3. "a" = 1. "r" the 18th letter, 1 + 8 = 9. "y' the 25th letter, 2 + 5 = 7. The total of all the letters added together, 1 + 1 + 5 + 3 + 1 + 9 + 7 = 27. Thus January = 27. 27 + 17 (the date) = 44. 44 + 4 (the hour) = 48. 48 + 3 + 1 (the minutes) = 52. 52 + 1 (for "a," of a.m.) = 53. 53 + 13 (for "m," the 13th letter of the alphabet) = 66.

The original news release stated the power of the earthquake to be 6.6 on the Richter Scale. The numerological total for the date and time of the quake combined with the numbers for its power was "6666."

Was that a coincidence? I wondered. Coincidence or not, I wasn't surprised.

One year later on January 16th, at 5:46 a.m., the Kobe earthquake struck Japan. I decided to try numerology on the day and time of that quake.

January equaled 27. 27 (January)+ 16 (the date)= 43. 43 + 5 + 4 + 6 (the time)= 58. 58 + 1 (for "a" of am.) = 59. 59 + 13 (for "m"of am.) = 72. The power of the Kobe quake on the Richter Scale was 7.2. Together the total for the date and the power was "7272." Then, I heard on the news that the Kobe quake, which occurred in 1995, was the worst natural disaster in Japan since 1923. That was 72 years earlier. Now I had three 72's. 7 + 2 = 9. Thus the Kobe quake was 999, which upside down is "666."

I also subtracted the numbers for the date and power of the Northridge quake 6666 from the numbers for the date and power of the Kobe quake, 7272. "7272" – "6666" = 606. I then realized that 606 is the common denominator of both quakes. 606 x 11 = 6666. 606 x 12 = 7272.

Coincidence or not, I found those numbers to be very interesting. Are we being warned?

CHAPTER 45

In early February of 1994, my parents come from Illinois for a two-week visit. I had prayed they would come to California at least one more time. I regretted the things that had happened on their last trip, and I wanted a chance to make up for it.

From the time they arrived in California, we were on the go. Mom and Gus had never been to Las Vegas, so I made reservations for three days at the Mirage Hotel and surprised them with the trip.

On our drive to Las Vegas, near the city of Barstow, California, the freeway was shut down due to a major auto accident. There was no way to go forward or turn around. We were stuck in the middle of a line of cars that had to be fifty miles long.

People patiently sat in their cars with the engines off, waiting for traffic to start moving. However, several hours later they encountered a problem. They had to go to the restroom and they couldn't wait any longer. A few people got out their cars, ran into the hills, and relieved themselves behind the bushes. Desert bushes aren't big or thick, so we could easily see what people were doing. Within minutes there were more people in the hills than were in cars. Everyone was laughing because they were in a dilemma, and they couldn't be concerned about modesty or pride. At least they knew, when the freeway reopened, they would never have to face one another again.

We were among the fortunate ones. Even though we had to squirm a few times, we managed to hold out until the traffic started to move. But before we arrived in Barstow we looked like we were dancing inside the car, and we wished we had joined everyone else in the bushes. By the time we arrived at a service station we were laughing so hard we barely made it to the restroom. If we had to have gone ten feet further, we wouldn't have made it.

When we arrived in Las Vegas my parents gambled a little. But the lights and shows mesmerized them. The highlight of their trip was seeing Siegfried and Roy. I was overjoyed that we all had such a good time.

When we returned to Ontario, we rested for one day, then started driving up the California coast to Hearst Castle in San Simeon.

On the way, we stopped in Northridge to see the damage caused by the earthquake. We saw the apartment building and the mall that had collapsed. The damage in my home area was nothing compared to the devastation we saw in the San Fernando Valley.

When we arrived at San Simeon we went on the tour of Hearst Castle. My parents thought the main house was a little ostentatious but its beauty overwhelmed them. The guesthouses, the grounds, and the swimming pool left them in awe. It was hard for them to believe that a family had actually lived in such a luxurious place.

The real highlight of that trip came after we left San Simeon: San Luis Obispo. Gus had been stationed at the army camp there during World War II, and he had never been back. To our surprise, when we arrived at the entrance gate, the guard gave us permission to drive onto the grounds. Gus was thrilled beyond words. As we drove around the camp, he even found his old barracks building. I could almost see his memories flashing through his mind.

Gus had a smile on his face and his eyes beamed with joy, but soon his sentimental memories touched his heart and he cried. My mother and I hadn't realized that seeing his old camp would mean so much to him. We were touched too. By the time we passed through the exit gate we were all teary eyed.

I don't know where the time went during the balance of those two weeks. It seemed as if my parents had just arrived and it was time for them to fly home. The day I took them to the airport I felt a profound sadness inside. I watched the plane disappear into the clouds before I left the airport, then drove home feeling as if a part of me was gone. I thanked God for the two weeks they had spent with me. I had prayed for the chance to treat them like a King and Queen and my prayers had been answered. That was the best vacation we ever had.

CHAPTER 46

Shortly after my parents left, I called Karen Kramer. She was quite distraught. The Northridge quake had caused major damage to her house. She felt the Devil was working double duty on her, and she couldn't cope with more problems.

I asked her if she would like me to arrange for the priest who had blessed my house to bless hers.

"Yes," she responded. "I would love for you to do that. All these bad things have to stop."

I called the priest, he consented, and a couple of weeks later I drove him to Karen's home. He gave her a bottle of holy water and blessed her house. Again, there were no wailing voices or furnishings flying around the house. Everything remained peaceful and calm. When he finished blessing the house Karen appeared to be appreciative, but I felt something wasn't quite right. I could tell she didn't want us to stay any longer. She seemed to be very anxious. Her behavior puzzled me, but I had things to do so I took the priest home and didn't think any more about it.

The next time I talked to Karen, she told me things were worse. She felt that having her home blessed caused the Devil to create even more problems. She was unhappy that I had brought the priest there. She didn't realize it, but she had fallen into the Devil's trap. He had purposely made her life miserable so she would be afraid of him. And afraid of him, she was. She couldn't see, or didn't care, that her fear of the Devil was letting him win.

"The Devil will never stop attacking you until he has no chance of winning," I said to her. "Bad things don't happen to evil people. He already owns their souls. Bad things only happen to people who fight the Devil and maintain their faith in God. He knows when people are weak and vulnerable, and as long as he has a chance to win, he won't give up. You can't give in to him no matter what he does to you."

But that didn't matter to Karen. All she wanted was to get the Devil off her back, and she didn't care what she had to do to accomplish it.

I talked with Karen on the phone several times after that day. Then without any explanation she no longer returned my calls. The last time I called her, her husband Stanley answered.

"This is George Newberry," I said to him. "I would like to speak to Karen, please."

"Hold on," he replied.

"George Newberry wants to talk to you," I overheard him say to Karen.

"Tell him I'm not at home," she replied. "I don't want to talk to him."

Karen was unaware that I heard every word she said. I was extremely disappointed to know she would do something like that to me.

"She isn't here," Stanley said to me.

To avoid a problem I didn't tell him I had overheard her comment.

"When she returns home will you have her return my call," I requested.

"Sure," he replied.

I decided to never call Karen again. And, she never returned my call. I still don't understand why she treated me with such disrespect. But there were two things I thought might have prompted her actions. One, she was interested in my script when there was a possibility of receiving money from investors. And, that didn't come to pass. Two, she could have thought the Devil was attacking her because she was trying to help me with my story. I hope that wasn't the reason. If it was, she gave the Devil just what he wanted and she wasted a year of my time.

I was disappointed, but I wasn't devastated by Karen's behavior. I had grown accustomed to dealing with people who were less than honest with me. Besides that, I had plenty of other duties to keep me occupied.

CHAPTER 47

In April of 1994, the lease was up on the Chino store. I decided to move the store to a larger location and add bridal gowns and accessories. Randy invested in the new store and he volunteered to manage it. That was a good opportunity for him and a perfect arrangement for me. It would allow me more free time to pursue my mission.

In July of 1994, the grand opening of Veils & Tails and Tuxedo Junction was held in Chino. The interior of the store was beautiful and, with Randy's expertise, business soon flourished. I was free to work on the screenplay and pursue the fulfillment of my dreams.

In the fall of 1994, I went with Jane Withers to the Gene Autry Museum Gala Ball. The event was held at the Century Plaza Hotel in Century City. At that function, Jane introduced me to Shirley Krims. Shirley had been the Director of Publicity for the Warner Brothers Facility in Burbank for the past twenty-two years. Jane told me to sit by Shirley and tell her my story. She hoped Shirley would be interested and help me find a producer for the script.

During the banquet I was able to tell Shirley only bits and pieces of my story, but she was fascinated and wanted to hear it all. That night I drove her home and we talked for several more hours.

Shirley was deeply intrigued by my story and believed in the things that I was trying to accomplish. She volunteered to put me in contact with someone who could produce the film.

I didn't know if Shirley was able to help me or not, but I had nothing to lose. The most important thing was that she had no doubt my mission was worthy of her backing.

About a month later Shirley introduced me to Antoinette, a lady whom she had met on the lot at Warner Studios. I told Antoinette everything that had evolved in my personal life as well as the basic plot for my story. She said she wanted to see the movie made, and she had the connections to get it produced.

My phone started ringing off the hook. Antoinette called me at least four times a week. I didn't know if she really had the contacts to get anything done, but at least she liked my story and seemed eager to help.

CHAPTER 48

In October of 1994, Marilyn Malmberg, the high school classmate I had called when I saw LaDonna floating over my bed in 1976, came to California for a two-week visit. Marilyn and I had been friends for 50 years. I was delighted that she was coming. My top priority for those two weeks was to see that she had a good time.

I have many close friends. And, many of those friendships developed when I was a child. Marilyn and I have been friends since we were four years old. We may be older, but our senses of humor have never changed. We always find time to joke and laugh.

"I'm going to fly into Ontario on Halloween night," Marilyn said.

"Then I'm going to pick you up at the airport in a costume," I replied. "Every Halloween night about 300 kids come to my door trick or treating, and I'm going to be dressed up in a costume this year. I'll have to wear it to the airport."

"Well, at least when I get off the plane I won't have any problem finding you," she replied.

"I should wear a costume to pick you up anyway," I chuckled. "After all, the Wicked Witch of the East will be flying into Ontario on Halloween night."

Halloween night I didn't dress in costume as I had planned. There were more trick-or-treaters at the house than I had expected and I was about 20 minutes late getting to the airport.

When I arrived at the airport a fully costumed witch with a long nose was sitting at the gate waiting for me. Held in her hand was a sign that stated, "The Wicked Witch of the East just flew in."

The moment I saw Marilyn I laughed hysterically.

"George," she said frantically, "you told me you were going to meet me here in a costume. I thought I would dress up like a witch and surprise you. When I walked off the plane dressed like this and you weren't here, I almost died. You should have seen everyone staring and laughing at me. I wouldn't have dressed up like this if I'd known you weren't going to be wearing a costume."

I could see why people were staring and laughing at Marilyn. I was laughing hysterically myself. It isn't a daily occurrence that people go to the airport and see a fully dressed witch walk off the plane.

Marilyn was embarrassed, but we have had a hundred laughs over that evening. I wish all people had a sense of humor like that. If they did, the world would be a much happier place.

I wanted Marilyn to meet Jane Withers during her visit to California, so I made arrangements for the three of us to go to dinner. Marilyn and I drove to Jane's home in Sherman Oaks. Then the three of us went to the Portofino Restaurant on Ventura Blvd. Jane and I frequented Portofino because it wasn't far from her home, and the food was always very good.

After dinner and extensive conversation, we went back to Jane's home. By then, Marilyn and Jane were talking like old buddies. I was delighted that Marilyn was having a good time. I took several pictures of Marilyn with Jane, so Marilyn could share her memories with her family and friends in Illinois.

On the way home Marilyn raved about how wonderful she thought Jane was. I was pleased to hear that, but I wasn't surprised. Jane has many fans. She is an expert on how to win friends and influence people. She has a bubbly personality and she's attentive to everyone she meets. I have attended hundreds of events with Jane, and I've never seen her be the least bit rude to anyone. Even if she's late for an appointment, she still takes time to talk to her fans. She makes them feel they are as important to her as she is to them.

A day or so later I received a call from Antoinette. She wanted Shirley and me to meet her at Lowe's Hotel in Santa Monica to discuss some developments on the movie project. I called Shirley. We made arrangements to meet Antoinette at the hotel the following evening.

The next morning Jane called. She wanted Marilyn and me to accompany her to a craft show in Santa Monica. I knew Marilyn would enjoy spending the day with Jane, but there was no way I could go. I couldn't cancel my meeting with Shirley and Antoinette. Thankfully, Randy came to the rescue. He made arrangements to take the day off and escort Marilyn and Jane to the craft show. Since they were going to be in Santa Monica, I invited them to join Shirley, Antoinette and me at Lowe's Hotel for dinner.

That evening I picked up Shirley at her home in the Hollywood Hills, and we drove on to Santa Monica. When we arrived at the hotel, Antoinette was already there. A short time later, Marilyn, Jane and Randy arrived and we were soon seated in the dining room.

The entire evening Antoinette raved about the script and boasted about the connections she had to get it produced. She told us a producer friend of hers was going to read the script and the movie would be made. That was nice to hear, but I wasn't the least bit excited. I no longer believed everything I was told.

Randy left before the rest of us, so when the evening came to a conclusion Marilyn rode back home with me. She had experienced a wonderful, fun-filled day, and she was ecstatic that she had been able to spend so much time with Jane.

"I can't believe I am in California hearing all this movie talk and meeting celebrities," Marilyn said. "George, you are so lucky to have friends like Jane and Shirley, and have someone like Antoinette helping you with your movie. Aren't you excited?"

I really wasn't. I didn't feel that anything was really going to develop through Antoinette's efforts.

"No, I'm not that excited, Marilyn," I replied. "I got excited when Karen Kramer told me the movie was going to be made. Look at what happened. Nothing! All I got from that was a big disappointment and wasted time. There are so many phonies in this town, you can't believe everything you hear."

A day or so later, Antoinette called me again. She wanted to have another meeting. This time a representative for an investor was coming to town and she wanted us to meet her at the Beverly Wilshire Hotel.

I took Marilyn to the meeting with me. When we arrived at the hotel, Antoinette was waiting in the lounge. We joined her at her table and ordered drinks. The lady who represented the investor was late. Antoinette called the lady's hotel room several times, but there was no answer. After waiting for several hours we decided to leave and go to Jimmy's Restaurant on Little Santa Monica Blvd.

When we arrived at Jimmy's, Antoinette tried to call her friend again. This time the lady answered the phone. She told

Antoinette that she had been detained at another meeting, and she would join us at Jimmy's.

From the moment the lady arrived and introductions were made she and Antoinette talked endlessly about movies and funding. Antoinette told the lady at least 10 times that a big movie was going to be made from my script. The lady went on and on about her ability to get the funding to produce it. Marilyn was enthralled by their conversation. But I wasn't the least bit excited.

Later, while I was driving home, Marilyn and I discussed the things that had been said that evening.

"George, I can't believe you aren't thrilled," she said. "You have someone who's going to get your movie made. Why aren't you excited?"

"The things they said sounded good," I replied. "But I'm afraid to get excited."

"Why," Marilyn asked. "George, it's really going to happen! Your movie is really going to be made!"

"Marilyn, I've been told that many times before," I replied. "There are so many phonies in Hollywood, it's pathetic. I don't believe everything I hear anymore."

I wanted everything that had been discussed to be true, but I was afraid to believe it. I didn't want to end up disappointed again. However, Marilyn felt so positive about everything that her enthusiasm slowly convinced me, against my better judgment, to halfway believe it too. Marilyn and I sat up all night talking about it.

I was never a secretive person. I believed in being very open with my friends. If I was anticipating good things to happen in my life, all of my friends knew it. It never crossed my mind that I had reason to hide my excitement from them. They knew previous hopes of my movie being produced had fallen through. Now there was renewed hope that my dreams might come true. I had to share my excitement. Within a week most of my friends knew everything Antoinette had promised me.

CHAPTER 49

The balance of the time Marilyn spent in California went by like a streak of lightning. We made good use of every minute she was here and enjoyed every one of them. Marilyn had to leave California, but our time together was not going to end. I wanted to see my parents and family, so I accompanied Marilyn on her return flight to Illinois.

My parents were both in their late seventies and I knew they wouldn't be around forever. I wanted to spend as much time with them as I possibly could.

Marilyn and I arrived at the airport in St. Louis. When we walked into the terminal, Gus was at the gate waiting for us. I was happy to see him, but I wondered why he was alone. Usually Mom was with him.

"Where is Mom," I asked.

"She's waiting for us in the baggage department," he replied. "She didn't think she could walk all the way to the gate. Her legs have been giving her a hard time."

That was something I didn't like to hear. Mom had congestive heart failure, and poor circulation caused her legs to cramp when she walked too far, but that was the first time she didn't meet me at the gate.

The second we stepped into the lobby of the baggage department, Mom spotted us. She was sitting in a chair by the entrance. She was excited to see me and her face beamed with joy. I rushed to her side and gave her a long and loving hug and kiss.

"I couldn't wait for you to get here," she said. "The last few days of waiting seemed like a week."

"Well, I'm finally here," I replied. "I was just as anxious to see you."

Within a few minutes, we retrieved our luggage and started walking to the car. We weren't halfway to the parking lot before Mom got cramps in her legs and had to stop for a rest. She tried her best not to complain, but I could see that she was in agony.

I was heartbroken to see that her health was declining. It was all I could do to hold back tears. Facing reality wasn't easy. I could see that my parents were aging. My heart felt heavy, but I had to smile and think happy thoughts. We were together, and we were going to have a good time.

Gus always looked out for my mother's well-being. She couldn't have married a better man. He was concerned about her health and happiness and never thought of himself, but my heart also ached for him. I could see that Gus wasn't in the best of health either. I wanted to enjoy every minute I could be with them.

I did have a good time on that trip. However, there was one thing I didn't enjoy. They had purchased their tombstone, and they wanted to show it to me. I didn't want to hurt them, but I couldn't stand the thought of looking at it.

"I don't want to see it," I said.

"Why not?" My mother replied. "Sooner or later we all will need one. We wanted to have everything ready and paid for so you kids won't have to bother with anything after we're gone."

"Besides that," Gus added, "we got to pick it out ourselves."

A visualization of their tombstone flashed through my mind, and tears started pouring from my eyes.

"I would rather not see it," I said to them. "I don't even want to think about it."

"But wouldn't you rather have us with you when you see it than wait until we're gone?" Mom asked. "We're proud of it."

The more I thought about it, the more I realized Mom was right. If I saw it for the first time after one of them was gone, I would hurt even more.

"Ok," I replied. "I'll look at it under one condition."

"What's the condition," Gus asked.

"I will go with you if you let me take a picture of both of you standing behind the tombstone making funny faces at the camera. Then when I look at the photo, it won't seem so serious."

They agreed to my condition, and I took the picture. Every time I look at it, the expression on their faces makes me feel a little better.

CHAPTER 50

While I was in Illinois, I received many phone calls from Antoinette. Her persistence and apparent dedication to having the film produced further convinced me that she was telling the truth. According to her, she had a producer committed to making the movie. She always called me from her cell phone and talked endlessly. I couldn't believe she would pay the high cost of those calls unless she was truly sincere. However, I did think that perhaps someone else paid the phone bill.

When I returned to California, by a total fluke, I met someone who knew Antoinette personally. Antoinette had told Shirley and me that she owned a fleet of antique cars that she rented to the studios. Plus, she owned a fleet of boats that were docked at Marina Del Rey. Now, I had unexpectedly met someone who knew the truth. He knew all about Antoinette.

According to him, Antoinette knew the owner of the cars. But, she didn't own one of them. Plus, she didn't own a fleet of boats, nor have the connections to get a movie made. Shirley and I had sensed something was wrong with Antoinette's story. Now at last, we knew our suspicions were right.

From that moment Shirley and I stopped communicating with Antoinette. The calls from her gradually came further and further apart. Eventually, we never heard from her again.

With the Antoinette episode behind us, Shirley wanted to see the videotape of the television show I had appeared on in Raleigh, North Carolina. Shortly before Christmas I went to her house to show her the tape. We only watched a few seconds of the tape when it abruptly stopped. I immediately got up to see what the problem was. I pushed the stop button and the play button, but nothing happened. I pushed the rewind button. Still nothing happened. I pushed the eject button and discovered the tape was jammed inside the VCR.

After thirty minutes of pushing buttons and tugging on the cassette I was able to free it, but the tape was still entangled inside the VCR. I carefully pulled on the tape. Inch by inch it slowly came out of the machine. A few feet of the tape were mangled, but at least the tape wasn't broken. I wound the tape back onto the cassette then inserted it back into the VCR. I pushed the rewind button and the play button but the machine didn't respond. I realized the machine had not only eaten up my tape, the VCR itself was broken.

Shirley wasn't happy that her VCR needed repair, and I wasn't happy that my tape was mangled, but we both managed to joke about what had happened.

"The Devil didn't want me to see that tape," she chuckled. "This is very strange. I've never had a problem with my VCR."

"I do believe the Devil had his hand in this," I replied. "He's relentless. If there's any way he can mess things up, he's going to do it."

I tried to get the VCR to work but my efforts were futile. It had to be repaired. I knew it would be difficult for Shirley to take it to a repair shop, so I volunteered to take it for her.

I drove home with my mangled videotape and Shirley's broken VCR. We had intended to spend a peaceful evening watching the videotape and discussing the television show, but it turned out to be an evening filled with frustration. The only thing we discussed was the Devil.

The next morning I took Shirley's VCR to a repair shop in Ontario. The repairman diagnosed the problem and told me I could pick it up the following day.

Shortly after I returned home from the repair shop, the doorbell rang. I opened the door and was surprised to see Edie Boudreau. Edie had assisted me when I tried to write the book in 1983. She was now an editor of a business magazine. We rarely had time to get together, but we talked on the phone quite often. We had developed a close friendship during the past 11 years. Plus, Edie was well aware of all the strange experiences I had encountered.

I invited Edie into my home. We walked through the living room, stepped down into the family room, and sat on a sofa facing the TV. I explained to her what had happened to my tape and Shirley's VCR the evening before.

"Is that the only videotape you have of your shows," Edie asked.

"Yes, it is," I answered.

"You shouldn't use your master tape," she replied. "If it gets ruined and it's your only copy, you're out of luck. If you'd like, I can make you a copy, then you could use it and save your master. If the copy gets damaged, I can always make another one."

I had never thought about making a copy of the tape. My original could have been destroyed. Then I would have had no proof that I had ever been on a television show. It was my good fortune that Edie stopped to visit that afternoon. When she left for home she took my master tape with her.

The following day Edie returned my original tape along with a copy. That afternoon I picked up the VCR from the repair shop and headed back to Shirley's house. Shortly after I arrived at Shirley's, I had the VCR hooked up. Once more we were ready to watch the tape.

I pushed the play button and sat in a chair next to Shirley. The moment the tape started to play I noticed something was wrong. There were noises and static in the soundtrack that hadn't been there before.

"Boy, the connections on Edie's VCR must be bad. This tape isn't fit to send to anyone," I said to Shirley.

"Well, at least we can see the picture this time," she chuckled.

The original soundtrack was audible so we tried to ignore the noises and watch the show. For a moment the noises stopped, then they started again and began to get worse. A repairman was working on Shirley's roof and she thought the noises might have been caused by some tools he was using. I turned the VCR off and the noise ceased. We knew the problem was in the soundtrack.

I pushed the play button and we continued to watch the show. The erratic noises were still there. There was static, clicking, and every so often what sounded like the chirp of a bird. The noises were irritating, but at least we were able to hear the original soundtrack.

We were watching the segment where I demonstrated the numerology on the blackboard when the noises started again. This time there was a bizarre addition to the sounds. A strange voice repeated "I-O, I-O." Shirley and I looked at one another with puzzled expressions.

"Gee, this is really weird," Shirley said. "What's going on here?"

"I have no clue," I replied. "But I agree with you. It is weird."

On screen I had just added up all six columns of numbers that totaled 1998. I had also just said to Dea, the show's hostess, "I was trying to figure out what 1998 meant, then the thought popped into my head that 1998 would be the third time something would occur since Christ died."

On the videotape, I started to divide 3 into 1998 to get the resulting 666. The second I placed the three on the blackboard a raspy, deep, evil voice emitted from the television. "Lies, Lies, Lies," the chilling voice uttered loud and clear.

Neither Shirley nor I could believe what we had heard. I thought Shirley's TV had picked up a few seconds of soundtrack from a television show. So, I rewound the tape and played it again. We were amazed to hear the same creepy voice repeat, "Lies, Lies, Lies."

This time I let the tape keep playing. Less than a minute after we heard "Lies, Lies, Lies," the same voice mumbled something else that wasn't as audible. I rewound the tape and played it over again and again. We understood the voice to say, "My daughter learned how to speak Spanish."

Shirley was almost beside herself. The first time we tried to watch the tape, her VCR ate it up and her VCR broke. Now, it seemed to be hexed by those noises, and that evil voice.

"George, this is really scary. Maybe the Devil does have a hand in all of this," Shirley said.

"I have no explanation for it," I replied. "It is either an act of the Devil, or someone is playing a practical joke on me. If it's a joke, I have to find out now."

I immediately called Edie on the telephone. When she answered, I didn't even take time to say hello.

"Edie, are you trying to pull some kind of joke on me?" I asked.

"What are you talking about?" she questioned.

"There are all kinds of noises on that tape you made for me. And there's a creepy, evil voice that says, 'Lies, Lies, Lies.'"

"Come on, you've got to be kidding me," she exclaimed. "There couldn't be anything on that tape. All I did was put your tape in one VCR and a blank tape in the other. I made two copies so

I would have one to watch later, and I didn't even turn the television on when I made either one of them."

"Then something weird is going on," I said. "Are you sure one of your kids didn't do anything to that tape?"

"My kids weren't even in the house at any time that tape was here," Edie quickly responded. "No one touched that tape but me."

"I'll be home around 11:30 tonight," I replied. "If you want to see and hear it for yourself meet me at my house."

"I don't know how anything could've happened to that tape," Edie responded, "but I'll meet you at 11:30. If I get there before you do I'll wait in front of your house."

I was a little concerned that Shirley might be afraid to stay alone after all that had occurred; especially after Edie indicated that it was not a joke. But, Shirley was more amazed that it had happened than she was frightened by it.

That occurrence was mind-boggling, but we managed to find some humor in it. By the time I left Shirley's house, neither of us was the least bit frightened.

When I arrived at home, Edie was already waiting in front of the house. As we started walking toward the house, I pointed at the tape I held in my hand. "You're never going to believe what's on this tape," I said.

"While I was waiting for you to come home, I watched the tape I kept for myself," Edie replied. "There were no noises or voices on it. It's kind of eerie that I made two copies and at random picked one of them for you, and it's the one with voices on it."

I couldn't have agreed more, but I still thought that I might find out it was a joke. When we entered the house we immediately went to the family room and I inserted the tape into the VCR. Edie and I sat on the sofa and intently watched it. When Edie heard the noises in the soundtrack she didn't seem to be too concerned, but when she heard the voices it was another story.

"Lies, Lies, Lies." The evil voice uttered once more.

Edie gave me a funny look. "That's weird," she said.

"Wait a minute," I replied. "In a few seconds you'll hear more."

About thirty seconds later the same voice mumbled, "My daughter learned how to speak Spanish."

"Oh my gosh," Edie instantly remarked. "I think I know what happened. I think that's my neighbor talking. I have a neighbor down the street who is a drug dealing low life, and he has a CB radio. I think the noises are from his radio, and that voice sounds like his!"

"Are you sure no one messed with that tape," I questioned. "Do you think it's possible your son did it for a joke?"

"No one was at my house while I had that tape, and I know no one touched it besides me," Edie replied. "I'm almost positive that's my neighbor's voice. Let me take the tape home with me. My son will know if it's him or not."

Edie took the tape home with her that night. The following day her son confirmed that the voice on the tape was that of her neighbor. Edie and I were both dumbfounded. Her neighbor had no way of knowing that Edie was making a copy of my tape. For some strange reason her VCR had picked up and recorded his CB Radio broadcast.

The buzzing, static and clicking sounds had emitted from Edie's neighbor's CB radio. But, he had not been conversing with another person. If he had been, that person's voice was not audible on the tape. It was uncanny that the only words the neighbor had said loud and clear were "Lies, Lies, Lies."

Why did he only say those words? Why did he say them at such a pertinent moment? Why did it even happen? Neither Edie nor I can answer those questions. However, if it ever becomes necessary, we will both take polygraph tests to verify that neither of us tampered with, or have any knowledge of anyone else tampering with, that tape.

Edie's neighbor was eventually arrested on a drug charge and served time in prison. Unless her son told him that his voice appeared on that tape, he is still unaware that it happened.

I'm sure some people find it hard to believe that such a thing could have happened. When I first told my friend Orland about it, he didn't believe me. He listened to the tape then scowled at me.

"Come on," he said. "Now you're going a little too far. What are you trying to pull here?"

"I'm not trying to pull anything." I exclaimed. "I'm telling you the truth. It really did happen."

"I don't believe it," he replied. "You messed with that tape. You'd better not try to get anyone else to believe a story like this, or you're going to lose all of your credibility."

I was shocked that Orland thought I was lying to him.

"That's probably one reason it happened," I responded. "I know the Devil caused those voices to be on that tape. He wants to scare me so I will stop working against him. He knows that isn't going to happen, so he wants to destroy my credibility. I don't care what people think. I'm not going to deny the truth just to keep people from thinking I'm crazy. The Devil is never going to shut me up."

Eventually, Orland heard the same story from Edie. Then he knew I had told him the truth.

CHAPTER 51

In April of 1995, there was a Spring Bridal Market at the Hacienda Hotel in Las Vegas. Randy and I went to that show to order new inventory for the store.

We left for Las Vegas in the early afternoon of April 11. We were about halfway there when I decided to call my parents in Illinois. I dialed their phone number on my cell phone. My stepfather answered. He usually talked for a couple of minutes and then passed the phone to my mother. This time, for some reason, he stayed on the phone and kept talking. When my mother got on the extension, he stayed on the line and remained a part of the conversation for the duration of the call.

"That was unusual for Gus to talk for such a long time," I said to Randy. "He usually doesn't have that much to say."

We arrived in Las Vegas and checked into our room at the Hacienda Hotel. That evening we went to dinner with friends who were also attending the bridal show. Then, in anticipation of a busy day to follow, we went to our room and got a good night's sleep.

The next day we were walking around the showroom when I heard a page being repeated over and over on the public address system.

"George Newberry, please go to a house phone. You have an urgent message."

My heart started to pound and I immediately searched for a house phone. I was afraid to find out what the emergency was. I was hoping it was only a problem at one of the stores.

I found a house phone and I immediately called the hotel office.

"This is George Newberry," I said. "I understand I have an urgent message."

"Yes, Mr. Newberry," the operator answered. "You have a message to call your brother, Tom, as soon as possible."

I knew then that something terrible must have happened. Tom wouldn't be calling me in Las Vegas for any trivial reason. My heart sank to the floor. I thought something had happened to my mother.

Randy and my friends were with me when I made that call. When Tom answered the phone, I feared what he was going to say. I started to cry before he even told me what had happened.

"Hi George," he said. "I have some bad news."

My legs felt weak. I just knew he was going to tell me that Mom was in the hospital, or even worse, dead.

"What's wrong, Tom," I asked.

"Dad is dead," he answered.

Those words hit me like a sledgehammer. It was the last thing I had expected to hear. Gus had never been sick a day in his life. Mom was the one we had been concerned about. In my wildest imagination, I never thought Gus would die before she did.

For a moment I was in shock, then I managed to regain my senses.

"What happened?" I asked.

"He had a massive heart attack," Tom replied. "He died instantly. He had breakfast with Mom, then he went into the bathroom. She heard a crash and went to see what happened. She found him crumpled up on the floor. She called the paramedics, but there was nothing they could do. They said he was dead before he hit the floor."

I was so distraught I could hardly talk. I managed to tell Tom I would be there as soon as I could schedule a flight. Then I hung up the phone and cried.

My friends knew I had received some horrible news and they tried to comfort me. I controlled my emotions enough to tell them what had happened, then I went to my hotel room and made reservations for the flight to Illinois.

Ever since I started writing this story, I have prayed that my mother would live to see good things come from it. She had major heart problems and I was very concerned about her health. I never thought for a moment my stepfather was the one who wouldn't be around much longer. I took it for granted he would live to be a hundred years old.

Attending Gus's funeral was one of the hardest things I have ever done in my life. I had never lost anyone who was close to me. When my real father died I was only four years old, and my young mind didn't comprehend what had happened. I didn't experience the pain and anguish my mother endured.

For the first time, as an adult, I had to cope with the death of an immediate member of my family...my stepfather, who I loved very much. I felt the pain of his loss and the pain of seeing my mother so devastated. She had lost a husband who had been her constant companion for 50 years. For her sake, I had to suppress my emotions and give her my support.

The first time my song, "Those Were the Times to Remember," was played for an audience was at Gus's funeral. The music played and the lyrics echoed throughout the funeral home. I intently listened to every word. My heart was heavy as I listened, but I was proud I had written such a beautiful song. The lyrics befit my mother and anyone else who had lost a loved one. Sobs and sniffles from others let me know that my song touched many people. The lyrics are as follows.

<u>Those Were the Times to Remember</u>

I remember the arms that held me close,
Hearts filled with love so true.
I remember the times we used to share,
Things that we used to do.

I remember the plans we used to make,
Words only eyes could say.
I remember the touch that made us one,
Games that we used to play.

The times of joy and laughter
Fill our lives ever after.
Sunshine still is our master,
Even if rain should fall.

Yesterday's tears and sorrows
Can't destroy our tomorrows
Just memories of good times we'll borrow,
For those were the times to remember.

I remember the good times shared with you,
Though you've been gone for a while.
I dream of the time we'll meet again,
Then once more you'll make me smile.
For those were the times to remember.

When the song ended there was hardly a dry eye in the funeral home. After the service, almost everyone told me they loved it. Their comments of approval were good for my self-esteem, and they made my mother very proud.

A few days after the funeral I had to return to California. Mom was attempting to cope with Gus's death, but her road to recovery was not going to be an easy one. For the first time in her life, she was going to be living alone. I knew my siblings would give her their constant support and she was being left in loving hands, but getting on the plane and leaving her behind wasn't easy. I flew home with a heavy heart.

CHAPTER 52

During the summer of 1995, Shirley contacted someone who she thought would be the perfect candidate to produce the movie. His name was Robert Munger. Bob, as she called him, had the original idea that was the basis for the film, "The Omen." My story was a cross between "The Exorcist," "The Omen," and "Ghost." Shirley thought that since the subject matter was similar to that of "The Omen," Bob would find it interesting. Plus, she knew him quite well. At one time, Bob had an office next to Shirley's office on the Warner Bros. lot.

Shirley called Bob and told him about my story. She aroused his interest and persuaded him to meet with us and discuss it. The meeting was scheduled for September 15, at the Hyatt Westlake Hotel in Westlake Village.

The day of the meeting I drove from Ontario to Shirley's house on Mulholland Drive, then we headed for Westlake Village. We were already seated in the hotel dining room when Bob arrived.

I was surprised when I met Bob; he didn't look anything like I had expected. He was a thin, lanky, older man in casual dress. Not that it made any difference, but I had expected to meet someone more sophisticated and professional looking.

We made our introductions, then Bob sat down and we ordered lunch. I was anxious to talk about my script, but it was obvious that Bob was in no hurry to listen. He did most of the talking. I did most of the listening.

I felt very uneasy. Bob wasn't a warm person and he wasn't easy to talk to. He talked about "The Omen" and a film he had made while at Warner Bros. called "Born Again," which starred Dean Jones. He also made sure that I was aware of the fact that he was a "Born Again Christian."

I felt a little more at ease after he told me how religious he was. I figured that if he was religious and a good Christian I could surely trust him.

Bob told me about his friendship with Shirley and how he thought so much of her.

"You're really lucky to know this lady," he said to me. "I really didn't have time to come here today, but I took the time just because I wanted to see Shirley."

I felt a twinge of anxiety when he said that. He was really saying that he wasn't there because he had any interest in my script, or me. But that didn't matter. I knew that, if for no reason other than showing his respect for Shirley, he would listen to my story.

After lunch he finally got on the subject of my script. "Shirley told me you have a very interesting story to tell," he said.

"It's not only interesting, it's a true story that's hard to believe," Shirley added. "I think when you hear what it's about, you'll know why I thought you'd be the right person to produce the movie."

"Well," Bob said as he looked at me, "What's this script about?"

"Do you want me to tell you a brief synopsis of the story, then you can read the script later?" I asked.

"No," he replied. "Tell me everything. Start at the beginning and tell me the entire story just as if you were giving a book report."

Well, I did just that. By the time I finished talking, Bob knew everything about my script and just as much about my life. I don't think I missed telling him one thing. I showed him evidence to validate my story, and gave him names of people who would testify that it was true.

Bob knew that I believed I had to fulfill a God-given destiny. And, he knew I was on a mission to help prove the forces of good and evil exist. I thought since Bob was such a devout "Born Again Christian," he would be enthralled by my story and want to see that it was told. I didn't expect him to believe everything I told him, but I did expect him to have enough interest to be inquisitive. Was I ever surprised. He didn't ask one question about anything, including the supernatural things that had happened to me. His only interest was whether or not it was politically correct with his religious beliefs.

"Well, Shirley couldn't have come to me with this script at a better time," he said. "I just got word from my attorney that I have unlimited funds to do any project I want."

Regardless of what he thought about everything else, Shirley and I were excited to hear him make that statement. We assumed it was an indication that he had some interest in the script. Otherwise, he would not have told us he would have access to the funds to produce it.

I left a copy of the script, which at that time was titled "False Prophets," with Bob. He promised Shirley that he would meet with us again as soon as he had read it.

"I'll read it, and we'll see where we go from there," he said. "Even if it isn't right for me, I know people who may be interested."

I drove home from that meeting with mixed emotions. I didn't feel that the vibes coming from Bob were that good. However, I had just met him, and I didn't want to make a hasty judgment based on my first impression. From past experience, I knew that first impressions of people aren't always right. Besides that, Shirley held the highest esteem for both Bob and his character, and I had the greatest respect for Shirley and her opinions. I knew many of Shirley's friends, and I knew her character was beyond reproach. If Bob was not the devoutly good Christian he professed himself to be, Shirley was not aware of it.

For several months Shirley and I patiently awaited a call from Bob. It never came. We knew it didn't take months to read a script, and we didn't want to waste more time if he wasn't interested, so in December of 1995, Shirley called Bob. As soon as her conversation with him concluded, she called me.

"Well, I just talked to Bob," she said. "He didn't say that he had read the script, but he told me there were several people he wanted to talk to about it. He didn't want to ask them to read the entire script, so he wants you to write a treatment (a synopsis) of the story. When you're finished with the treatment he will meet with us again."

"I've never written a treatment," I replied. "How long should it be, and how much story detail should be in it"

"I can loan you a treatment that Milton wrote for one of his scripts," Shirley replied. "Then you can get an idea of what you need to do."

Shirley's deceased husband, Milton Krims, had been one of the top screenwriters in Hollywood. Among his greatest successes were "The Sisters," with Bette Davis; "Confessions of a Nazi Spy;"

and "Iron Curtain." I felt fortunate that Shirley would loan me one of his treatments. I couldn't learn from a better writer and I appreciated that very much.

Shirley loaned me Milton's treatment and I read it over and over. I learned enough from his example to write my own treatment, but I lacked the enthusiasm I needed to start writing. Months passed, and I still hadn't started to work on it. Writing that treatment was not one of my priorities, and I was never the least bit concerned that Bob was waiting for it.

CHAPTER 53

Almost a full year slipped by before I knew it. Business had been good at both stores, but problems continued with the apartments. Tenants would not pay their rent and every tenant I evicted thrashed their apartment before they moved out. I not only lost the rental income, but I had to pay the eviction fees and the cost of repairs. At least the stores were doing fine and I was able to pay the losses without going further into debt. That's probably why the time passed by so fast. Even though I had financial problems, I was able to handle them.

With the blink of an eye, it was January of 1997. Business was still good. I was still able to pay all the bills, and there weren't a lot of problems to stress over. The only thing that bothered me was that my procrastination had caused me to waste a lot of time that I should have spent writing.

It had been one year since Bob Munger had requested the treatment on my script, and I hadn't written the first page. The procrastination had to stop. I was never going to fulfill my mission unless I put more effort into it.

Once I started to write, I became more enthusiastic. The right words came easy and within a couple of weeks the treatment was finished. At long last, I asked Shirley to call Bob and set up our next meeting. I was surprised she didn't go into shock. It had been a long time since Bob had requested the treatment. However, Shirley called him and he agreed to meet with us again on February 26, at the Hyatt Westlake Hotel.

At that meeting I felt a little more at ease around Bob. He reassured me of his good intentions and his desire to make the movie.

"I just received word that I am going to have access to hundreds of millions to fund my projects," he added.

That comment made me feel even better. I thought he must be interested, or he wouldn't have told us about the money that he was going to receive.

After lunch and a short conversation I gave the treatment to Bob, then Shirley and I headed for home. This time we were both excited. I finally believed that Bob was sincere, and there was a chance my story would be told.

CHAPTER 54

On April 11, Shirley, Bob and I again met at the Hyatt Westlake Hotel to discuss the treatment. Shirley and I thought that Bob was going to tell us he was ready to proceed with the project. As soon as we finished lunch and he started discussing the script, we knew that was not so.

"I don't need the treatment," Bob informed me. "You need to have a complete rewrite done on the script."

I was both puzzled and disappointed. I didn't understand why he didn't tell me the script needed a complete rewrite in the first place. I tried not to show it, but I was very irritated. I had wasted a year of my time over a treatment he didn't even need. He should have known the script needed a rewrite before he asked for the treatment.

I saw no reason the script had to be rewritten. I knew I didn't have a script that was ready to go into production, but I didn't think it needed a major overhaul before it could be sold.

"Can't it be sold the way it is, then rewritten later," I asked Bob.

"If you want me to produce this film, you have to have a rewrite done now. I can't do anything with it the way it is," he sternly answered.

I was very uneasy with what Bob had said. But, he was a professional filmmaker, and I had to trust him. Besides, he had me convinced that he wanted to make the movie. He was the only hope I had to get my story told.

"Nothing can happen with this script the way it is," he said. "You have a good story, but you're not a professional writer. It needs to have a complete rewrite done by a professional screenwriter. I can't do anything with it the way it is."

It was then that I wondered if he had really read the script. Why hadn't he told me that before? I knew I had a good story. He couldn't deny that. Granted, I wasn't a professional writer, but I had already paid a professional writer to work with me on the script. The script needed to be polished, but I couldn't see any reason for a complete rewrite.

I was uneasy with Bob's statement, but I kept quiet and listened to him. I wanted him to make my movie.

"A professional rewrite will cost about forty-eight thousand dollars," Bob said. "But if I work on the rewrite with a writer, I can save you a lot of money. I will work for free. Then I can probably find a good writer who will do it for twenty thousand dollars."

I looked at Bob for a moment without responding. He knew I didn't have twenty thousand dollars to pay for a rewrite, and he could see I was extremely disappointed.

"I don't have twenty thousand dollars, Bob," I responded. "I've lost so much money since I started working on this story that I have none left. I've been taken to the cleaners by everyone who has had a chance to cheat me."

"There is an alternative," he said. "You can have a writer do the rewrite on his own, but you will lose the rights to your story. If you pay for the rewrite, you will retain all rights to the script."

"If I pay for the rewrite, what happens to the script when it's completed?" I asked. "There are no guarantees that I will get my money back."

"Well," he said. "I hope you sell it to me. The finished script could be worth a million or more dollars. You will end up making more money from the script than anyone else."

I felt somewhat better hearing that comment. At least if I had the money to pay for the rewrite I knew I would get it back.

"Bob," I replied. "I would love to be able to pay for a rewrite, but I have been ripped off so many times that I don't have twenty thousand dollars, and I don't know where I can get it."

"You've had all kinds of miraculous things happen to you," he said. "All those things have given you a lot of faith in what you're doing. I haven't had any miracles happen to me. If God really wants this movie to be made, He will see that you get the twenty thousand dollars. If that happens, I will have my miracle. It will be a sign to me that I am to make this movie. Then I can be just as excited about it as you are."

Bob made a commitment to make the movie, and I wanted to do anything that was going to help get the movie made. I had a mission to accomplish and I couldn't let money stand in my way. Bob was serious about producing the film, but he wouldn't do it without a rewrite. One way or another I had to come up with the money.

"All right Bob," I said. "I have to do everything within my power to tell this story. No matter what I have to do, I will get the twenty thousand dollars. I have no choice."

That statement basically ended the meeting. We all said our good-byes and soon Shirley and I were on our journey home.

As I drove home, I though about the things that were said at that meeting. Twenty thousand dollars was the only thing that stood in the way of my movie going into production. I couldn't let money keep my story from being told and prevent me from fulfilling my mission.

"I have to get twenty thousand dollars," I said to Shirley. "If I must, I will sell twenty thousand dollars worth of my antiques."

"You shouldn't do that," Shirley responded. "I don't think you should give Bob twenty thousand dollars. I've never heard of a writer being charged for the rewrite on a script."

"Bob said he would make the movie if I paid for the rewrite," I replied. "If I don't come up with the money he won't do it. Maybe God is testing me, to see if money is more important to me than my mission. Maybe the Devil is trying to block the story from being told because he knows I don't have the money. Whatever the reason, fulfilling my mission is more important to me than my furniture."

Shirley agreed that maybe I was being tested, but she wasn't comfortable with the fact that Bob had asked me for the money.

"I still don't think you should give him any money," she advised.

"Bob said if God wanted him to make the movie, God would see that I got the money. Then he would have his miracle, and he would be just as excited about making the film as I am. Bob knows I don't have the money, and there is no way that I can borrow it. One way or another, I have to give him his miracle."

I took Shirley home and drove on to Ontario. When I pulled the car into the garage, it was about 10:30 p.m. The second I stepped into the house, the phone rang. I rushed through the darkness into the family room and picked up the receiver. It was a call from Mel, one of my Hollywood friends.

"Hey, George, how ya doing," Mel asked.

"Well," I replied. "Right now I am excited and depressed at the same time."

"What do you mean," he asked.

"I'm excited because a producer wants to make my movie. I'm depressed because I need to pay twenty thousand dollars for a rewrite or I will lose ownership of the script. I don't have that kind of money anymore. I've had too many losses. I think I'm going to sell some of my antiques. My story is more important to me. My furniture can always be replaced."

"Don't do anything yet," he said. "Meet with me at my restaurant Monday evening. I may be able to come up with a solution. Maybe I can help you get the money."

I couldn't believe what I was hearing. I hadn't asked anyone for the money. I had just walked into the house, the phone rang, and now someone was offering to help me raise the funds. Was this the miracle that Bob was asking for? I truly hoped so. I was so excited that I immediately called Shirley and told her what had happened.

"Bob Munger may have his miracle," I said. "I will know for sure on Monday evening."

That Monday evening I met Mel at a restaurant he managed in Hollywood. Mel and I sat in a booth and ordered something to eat. I was eager to find out if he was able to help me raise the twenty thousand dollars, but I didn't want to bring up the subject. I didn't wait long until he mentioned it himself.

"George, I have a big surprise for you," he said. "I have the twenty thousand dollars you need to get your screenplay rewritten."

I couldn't believe what I had heard. "You've got to be kidding me," I said. "Where did you get that kind of money?"

"I'm getting ready to close a big deal that I've been working on for some time," Mel replied. "I will clear enough money on it to give you the money and never miss it."

"But why would you do something like that for me," I asked. "People never do anything like this. I've never had anyone hand me one penny in my life."

"George," he said, "you have a dream. You've worked on that dream for a long time and you haven't given up. I want you to see your dream come true."

I was almost in a state of shock. I had never had a surprise like that in my life. I couldn't believe that someone cared enough about me to help me. I was so touched that I started to cry.

"Mel, you're part of a miracle." I replied. "Bob Munger said that if God wanted my movie to be made, He would create a miracle and see that I got the twenty thousand dollars. I didn't even ask anyone to help me, and now I'll have the money. That has to be a miracle."

"You can call it a miracle if you want," Mel responded, "but I think a lot of you and I want to see your movie made."

"But what happens if I can't pay you the money back," I asked.

"You get your script rewritten," he replied. "If the movie is made, you can pay me back. If the movie isn't made, you don't owe me one penny. Just seeing you happy is enough payment for me."

I couldn't believe something like that was actually happening to me. I hadn't even asked Mel to help me, and he was going to give me the money. No one, other than my parents, had ever offered to give me anything, let alone twenty thousand dollars.

I drove home from that meeting in a state of shock. I was afraid I was going to wake up and find out the whole thing was a dream. But it wasn't. This time something good had really happened to me.

On my way home from my meeting with Mel, I called Shirley from my cell phone and told her the news. She was elated that I was going to have the money without selling my furniture. She was happy and I was on cloud nine.

The next day Shirley called Bob and told him that he had his miracle. I would have the twenty thousand dollars to pay for a rewrite.

Bob immediately called me.

"Well, it sure didn't take long to get the money," Bob said.

"I can't believe it myself," I replied. "The whole thing has to be a miracle. I didn't even ask anyone to loan me the money and out of a clear blue sky someone offered to give it to me. Now you can be excited about making this movie, too."

"We may as well get things started," Bob replied. "I'll fax you a contract. Read it, sign it, and fax it back to me."

"Sounds good to me," I replied. "I'm ready to get things rolling too."

Bob immediately drew up a contract and faxed me a copy. In the contract there was a clause stating there were no guarantees the movie would be made. I was to pay $20,000 for a first draft script of my treatment and screenplay and I was to retain ownership of all copyrights. I had barely finished reading the contract when the phone rang.

"Hello, George," Bob said. "I called to explain the clause in the contract that states there is no guarantee the movie will be made. That is in there solely to protect me in case something beyond my control should happen, and I couldn't make the film. Don't pay any attention to it. Nothing is going to happen. I'm going to see that the film is made."

"But what happens to me if something should go wrong," I asked. "I have to pay back the twenty thousand dollars."

"Don't worry about it," he replied. "Nothing is going to go wrong. I'm the one that has to handle everything now. Just leave it up to me. You're going to make more money off this than anyone. By the way, now that we are going to be proceeding with the production of the film, you won't have to bring Shirley to any more meetings. There is really nothing she can help us with. She would just be in the way and slow us down. Also, don't show her the contract. This is strictly between you and me, and it is my job to look out for things from now on."

I hung up the phone and thought about what Bob had said. I questioned his motives. *Why would Bob not want Shirley to see the contract? Why should he care if Shirley attends the meetings? Is Bob trying to hide something Shirley may not like?*

Shirley didn't like the idea that Bob wanted twenty thousand dollars to pay for the rewrite. She felt it should be done after Bob had a commitment to have the film produced, not before.

I fretted for a while trying to decide whether or not I should sign the contract. I eventually decided to call Bob back. When he answered the phone, I explained to him that I was a little undecided about signing the contract. "I just want to make sure I don't make a mistake," I said.

"The only way you will ever see your film in production is to trust me," Bob replied. "I will never do anything but look out for your best interest."

Bob convinced me to place my trust in him. As soon as I hung up the phone, I signed the contract and faxed it back to him.

Bob must have sensed that I was still insecure. During the next few weeks he called me several times to reassure me that I could trust him. I wasn't at home when he made one of those calls so he left a message on my recorder. The message was almost identical to what he had told me several times before.

"Leave everything up to me and don't worry," he said. "I think the world of you, and I will never do anything but look out for your best interest."

Bob sounded sincere, and his message made me feel more secure about things. I played that tape over and over. As a matter of fact, I played it for most of my friends. They were even impressed that he cared so much for me. Since he was such a devout "Born Again Christian," we felt he meant every word he had said. I was finally convinced there would never be a reason I couldn't trust him.

The first meeting Bob and I had after I signed that contract was on May 22, 1997. It had taken Bob several weeks to retain a writer. Again, the meeting was held at the Hyatt Westlake. At that meeting Bob told me he had problems finding a writer. A writer had made a commitment to work on the script, but for some reason, was unable to fulfill his commitment.

Bob had already consulted another writer. The new writer, Tom Hubbell, had read my script and treatment and had given his critique to Bob.

"I was really surprised to hear Tom's comments," Bob said. "Usually when I have someone read a script they tell me the story isn't good or there's nothing there that could be turned into a commercial film. Tom really liked your story. He said your script has a lot of potential. He feels after a rewrite you'll have a script that could make a very commercial film."

I was happy to hear Tom's remarks, but I thought it was strange Bob wasn't able to see that for himself. I again wondered if

Bob had really read my script, but the trust he had convinced me to have in him kept me from worrying about it. Again, Bob made every effort to reassure me that his time spent on the script was donated solely to help me and get the movie into production.

Before I left for home I wrote a check payable to The Munger Organization for seven thousand dollars. The funds were to pay the first installment of the twenty thousand dollars that were to pay Tom.

I hadn't received any of the money that Mel had promised to give me. For some reason his business deal was temporarily delayed. But a surge in business at my stores made it possible for me to cover the funds.

On June 2, I received a fax from Bob. It was a copy of the critique that Tom had given on my script. Those notes were dated May 22, the same day Bob and I had last met.

Tom had not only read everything I had written—he had taken the time to subjectively and objectively analyze all of it. It was obvious he had put forth a lot of effort preparing his critique. He noted all the points he thought to be important concerning my Bio, my treatment, and my script. He also included research that he had done to back up each point.

The bottom line of Tom's analysis was that I had a very good idea and a good plot for a commercial film. He stated what he thought to be the weak points in the story and offered his suggestions of what it would take to remedy them.

I found one of the comments that Tom had written to be most interesting. "IF YOU ACCEPT THIS MISSION, MISTER NEWBERRY, HELL'S GOING TO THROW EVERYTHING IT HAS AT YOU."

Tom's remarks continued. "After George had his near-death experience and began to preach about the "Good News," his life went to hell. He was given his own personal 'Angel of Darkness' to torment him. Financial and personal setbacks were all around him. Saint Paul called this "A thorn in the flesh." George was being attacked. Yet, George overcame the roadblocks and torments that Satan threw his way. And because of his perseverance, George was rewarded with OBJECTIVE EVIDENCE of the work of the evil one. Evil was at work in the physical world around George. The search-warrant, cars not registered to him, a license plate 'LUCIFER.' George has copies of all these things. Other people may make the same claims, but they have no such physical evidence."

After I read that fax I knew Tom was the right person to help with the rewrite on my script. Tom believed in my mission and he was willing to help me fulfill it.

CHAPTER 55

On June 8, there was another meeting at the Westlake Hyatt. At that meeting, Bob brought Tom and we met for the first time. He didn't look anything like I had pictured him. I had expected to see someone who was about my size, 5'10" tall, 175 pounds, with brown hair. Tom was the total opposite. He was built like a lumberjack, and he had light blonde—almost white—hair.

Tom brought the beginning pages of his rewrite with him. I read them and I was happy to see he had left most of the scenes I had written basically intact.

The three of us then discussed the main plot points of my story. To my surprise, Bob suggested that some of the things I had written based on my true experiences should be removed from the script. I immediately disagreed.

"My story is based on true events," I said. "Those are the things that I experienced. It's important that those points stay in the script. They happened for a reason. They are the most important part of my story."

"But you don't need everything that happened to you included in the script," Bob remarked.

"My story has to have an influence on people's lives," I retorted. "If any of the things I've experienced are left out, the story won't have the same impact. Everything that happened to me has to happen to Amy. Those are the documented true things that give the story credibility."

Bob didn't seem to be too happy with my decision, but that was the way it had to be.

"I didn't borrow the money for this rewrite to change the story," I added. "My mission is to tell my story, not one that is made up by someone else. I agree that the story can be polished and enhanced, but the facts that it's based on can't be deleted or changed."

I felt confident with the work that Tom was doing. I wanted him to continue, and I wanted to work with him. But, Bob didn't want us to work together without him. As a matter of fact, he made sure that Tom and I didn't even have the chance to exchange phone numbers.

"How is Tom going to work on my script without consulting me," I asked Bob.

"I am the Story Consultant and I am the Producer. If you want to run the show, then you don't need me," he answered.

With that comment I felt that he really put me in my place. He was basically saying that if he didn't run things to his liking, he would quit. I was intimidated not only by his comment, but also by his arrogant attitude. I was afraid he would walk out and the movie would go down the tubes. I didn't want that to happen, so I shut up.

Shortly after that meeting, I gave Bob a second check for seven thousand dollars. There hadn't been that much progress on the script, but I didn't question why he needed more money so soon. Bob wasn't getting paid one cent, so I felt I had no reason to ask questions.

I had agreed to pay Tom twenty thousand dollars to work on my script, and I had to fulfill my obligation. Besides, I never once questioned Tom's credibility. I knew from the first time we met, that he was honest and he was the right person for the job.

After I gave Bob the second installment, the three of us had several more meetings. I can't remember the exact date each meeting was held, but my American Express statements document the dates of most of them. I picked up the lunch tab at almost every meeting we had. If I didn't pay the check, Tom did. Bob never offered to pick up a check.

Since he was working for nothing, I thought perhaps he assumed I should pay for his lunch. I have never been slow at picking up the check and it didn't matter that much to me. He was donating his time to save me money, so I thought the least I could do was pay for his lunch.

In early July of 1997, Bob, Tom, and I again had a meeting at the Westlake Hyatt. When Bob handed me the latest draft of the script, I noticed that my name had been deleted as a co-writer.

"Why is my name no longer listed as a co-writer," I asked Bob.

"When a script is rewritten the new writer's name goes on the script and the first writer's name comes off," he replied.

"But Bob," I exclaimed, "I checked with the Writer's Guild and, if I understand the rules, my name should stay on the script unless most of the things I wrote were replaced."

"Well," Bob responded, "we could always change enough to see that happen."

I didn't like his comment one bit. Again, I was seeing a side of Bob Munger I didn't like. I couldn't imagine why he would want to take my name off the script. I tried to hide the fact that I was angry, but this was one time I wasn't going to sit still and keep my mouth shut.

"I'm paying for this rewrite with the understanding that my name would stay on the script," I remarked.

"But you will still receive credit for the original story," Bob replied.

"I don't care," I quickly responded. "I want most of the scenes I wrote to stay in the script. I don't want my story destroyed."

"I don't care if George's name stays on the script," Tom exclaimed. "All I care about is getting it finished. George is trying to fulfill a mission. This is his story. I think his name should stay on it."

Tom's comments made me feel much better. But, Bob still didn't want to leave my name on the script.

"The script and all copyrights will still belong to you," Bob commented. "Whether your name is on it or not, you own it."

I wasn't the least bit happy with Bob's persistence to remove my name from the script. I didn't care what happened, I wasn't going to let him get by with it.

"That's not the point," I responded. "If my story is changed enough to remove my name, then it can't be the story I want to tell. I never agreed to pay for a rewrite that was going to ruin my story. I'm sorry, but that's the way it is."

"Well," Bob replied. "If it doesn't matter to Tom, then I guess it's ok."

I felt relieved that Bob dropped the subject, but I drove home from that meeting feeling uneasy. I knew Bob didn't care if my name was on my script or not.

I asked my neighbor, Rosemary Barnes, to go with me to my next meeting at the Westlake Hyatt. When we arrived at the hotel I introduced her to Bob and Tom. Tom was very cordial to Rosemary, but Bob was somewhat cold. I could tell he wasn't happy that I brought her with me. He gave her something to read and told her that she didn't have to be bored listening to our meeting. Thankfully,

Rosemary remained at our table and heard our conversation. As the meeting progressed, Bob became a little warmer toward her.

"When we first started with this project, I thought George's film was going to be one of my smallest projects," Bob said to Rosemary. "Now it looks like it's going to be my biggest."

I felt that Bob was sincere about making the film, or he wouldn't have made that comment to Rosemary. I was ecstatic. My movie was going to be made and it wasn't going to be a low-budget film. For the first time, I felt fortunate that Bob was going to be in charge of the production.

There is one change that we need to make," Bob exclaimed. "You have too many names in the script that start with 'A.' There is Amy, her mother Amelia, and her friend Anna. I think we should change the name of Amy to Christine."

"No, I won't agree to that," I said. "I don't want Amy's name changed. She is the main character in the story and that's been her name since I wrote the first page."

"But I think Christine is a much better name for the main character," Bob remarked. "It just sounds more like a Christian name than Amy."

"I will never agree to change Amy's name," I sternly remarked. "I may agree to change the name of another character, but not Amy's."

Bob wasn't happy that I was just as adamant about not changing Amy's name as I had been about the removal of my name from the script. But I didn't care.

I felt a wave of negativity pouring from Bob, and I questioned his motives and his credibility. It didn't seem to me that he was concerned about my story or my best interests. I was beginning to wonder if I could really trust him, and I was angered.

When he realized that he had raised my dander, he changed his attitude and backed off. He could see that I wasn't intimidated. He resented me contesting his judgment, but I didn't care. It was my story and my characters, and my money was paying the bills.

By the time that meeting ended things were much better. The air had cleared and there was no friction between us. His demeanor was pleasant, and as far as I was concerned it was time to put our issues to bed. We had to go forward. I knew that in the future we would again bump heads, but I couldn't let that affect our working relationship. There was no way we were going to agree on every issue. Perhaps the next time I would be the one who had to retreat.

I am glad Rosemary was at that meeting. She also sensed that Bob didn't want her there, but it was important to me that she witnessed what Bob had said.

At the next meeting, Tom gave me a draft of his latest revisions. That draft was dated July 16, 1997. The second I looked at it I was in shock. I instantly knew that Bob and I were headed for another confrontation. The name of Amy's mother had been changed from Amelia to Christine. I cringed when I saw it. The name, Christine, was no more right for Amelia than it was for Amy. I didn't like it. I didn't know why he was so determined to change the name of one of my characters to Christine, except that he thought it sounded so "Christian."

"I don't like the name change," I told Bob.

"Why not," he asked angrily. "You have to change the name of one of your characters. They can't all have names that start with, "A." It just won't work!"

"I don't want Amelia's name changed to Christine," I replied. "That name doesn't fit her. We can give her another name, but not Christine."

Amy and Amelia were character names that originated in the book I had started in 1983. I felt no different about changing their names than if I changed my own.

Bob remained adamant that Amelia's name should be changed. I could tell he was angry and I feared that if I didn't let him have his way he wouldn't make the film.

"Ok, let's change Amelia's name," I said reluctantly. "But let's come up with a name other than Christine."

"Why don't you like the name, Christine," Bob asked.

"I don't dislike the name," I replied. "I just don't feel it's right for Amelia. I personally don't see what's wrong with the names of Amelia or Amy. I think it would be a natural thing that someone named Amelia, would give her daughter a name like Amy."

"I don't agree," Bob responded. "One of the names has to go. Since you don't feel right with the name, "Christine," do you have any suggestions?"

I knew I was never going to change Bob's mind. I didn't want to change Amelia's name, but I also didn't want to have a major confrontation over it. Bob was too temperamental and I was afraid he would walk out on his commitment to make the movie. That was the last thing I wanted to happen.

"No," I replied. "At the moment, I can't think of another name that's befitting. For the time being, let's call her Suzanne. That's my daughter's middle name. We can always come up with another name and change it later."

"I don't see anything wrong with the name Christine," Bob replied. "But as long as Amelia's name is changed, I'll be satisfied."

Tom didn't care what the names of the characters were. His main concern was to get the rewrite finished so that I had a more commercial script. He knew what I was trying to accomplish, and he was doing his best to keep my original plot intact.

Tom and I never had a confrontation over changes he wanted to make. His ideas and efforts always seemed to enhance my original story. Tom was a professional writer, and I was happy with his work.

Bob was a horse of a different color. We were in constant disagreement. I never understood why he always wanted to make revisions that were going to wreck my story and destroy its purpose. I wondered if he had ulterior motives. But, I couldn't think of one reason that he would want to ruin my story. If the fruit of our efforts didn't produce a good commercial script, there would be no movie. He had as much to lose as I did.

We had barely agreed to change Amelia's name when I was hit with a second blow. Bob told Tom to delete more of my material from the script. He wasn't just trying to change the character's names, he wanted to wipe out some of the most important things in my story.

"We need to cut the scene with the Lucifer license plate," Bob remarked. "And all that numerology should go. No one is going to understand all those numbers."

"I don't agree," I responded. "The Lucifer license plate and the numbers are important parts of the story. All those things have to stay in the script. They're part of the things that make the story interesting. They give it credibility and they're things the audience needs to see."

I was adamant that my story was going to remain intact. Bob could never convince me to do otherwise. I couldn't allow him to destroy my story for the sake of making a movie.

"This script isn't based on fiction, Bob," Tom interjected. "I can see why George wants those things left in. They validate his story."

When I heard Tom's comments I breathed a sigh of relief. But I could see that Bob wasn't elated. He hadn't expected Tom to support me. Now he knew changing my story wasn't going to be an easy task. He expressed his disagreement and grumbled for a few minutes, but he didn't argue. Surprisingly, he changed the subject and the matter was dropped. I'm sure Tom's remark was the catalyst that prevented a confrontation.

I felt good things were accomplished at that meeting. I learned that Tom was definitely the right person to work on my script. And, Bob learned that I wouldn't agree to changes that would adversely affect my story, just because he dangled the promise to produce the film in my face.

CHAPTER 56

That meeting ended on a good note. We were leaving the hotel when I realized I had forgotten to extend an important invitation to Bob and Tom.

"My mother is coming from Illinois to spend the month of August with me," I quickly remarked. "Her eightieth birthday is on August the 8th, and I'm going to throw a party for her on the 10th. I would like for both of you to come."

Neither of them knew my mother, and it wasn't a short drive to my house, so I was pleasantly surprised when both of them indicated they wanted to attend. As a matter of fact, I was elated. I didn't think either of them would want to drive all the way to Ontario.

"I need directions to your house," Tom remarked. I knew, for whatever reason, Bob wanted to be the only bond between Tom and me, so I didn't dare write anything down and hand it to him.

"I'll fax directions to Bob then he can relay them to you," I replied.

Evidently that was the way Bob wanted it to be done. He never suggested that I fax Tom his own copy of the directions.

Shortly after that meeting my mother arrived in California, and my priorities changed. I wanted to spend as much time with her as I possibly could, and I intended to see that her birthday party was the best one she ever had.

Mom had been with me for only a few days when I received a call from Bob. He wanted to have another meeting on August 8th. I was busy secretly making arrangements for Mom's party but I agreed to meet with him.

The day before that meeting Bob sent me a fax. The message on the cover sheet read.

"George. I haven't been able to talk to Tom, so here is the uncorrected version including some bad typos. However, you will note lots of improvements and refinements. See you 12:30 P.M. at Hyatt Westlake Village tomorrow, Friday, August 8. Bob. Pages including cover page: 46.

Tom had completed 45 pages of the rewrite. It was dated July 31, 1997. I read it and I liked it. Tom had good ideas, he was a good writer, and he was staying with my story line. The only thing I was uncomfortable with was that Amelia's name had been changed to Susanne. I had intended for the name be spelled, Suzanne, with a "z." But an "s" or a "z" didn't make that much difference. No matter how it was spelled, it didn't spell Amelia.

The next day I met Bob and Tom at the Hyatt Westlake. Again, Bob told Tom to delete more of my material from the script. He wanted to cut more scenes that were based on fact. Once more I had to defend keeping those scenes in the new draft, and once again, Tom came to my defense.

"Bob," Tom remarked. "George has a commitment to fulfill a mission. It's his story based on his experiences. Anything less than that is defeating his purpose."

I could tell Bob wasn't pleased, but he gave up his efforts to delete more scenes from the script. He was a little disgruntled but he didn't seem to hold any animosity over our disagreement. When the meeting ended, I reminded them of the party and I made sure they had received directions to my house.

Bob and I were parked in the same section of the parking lot so we walked together to our cars. Before I got into my car he mentioned his plans for the movie.

"I would like to start getting this production put together," he stated. "Do you know anyone who can come up with five hundred thousand dollars?"

"For what?" I asked.

"If you can get someone to put up five hundred thousand dollars, I can get the production package put together. When my money comes in they will be the first to be paid back."

"I don't know anyone with that kind of money." I replied.

"Check around," he responded. "You may find someone who wants to make a good investment. And remember, I think the world of you, and I will never do anything but look out for your best interests."

"I'll see what I can do," I replied as I got into my car.

I drove home from that meeting filled with mixed emotions. I didn't know what to think about Bob. One minute he was trying to ruin my story and the next he wanted to get started on the production. And, he thought the world of me?

That evening I had a small family celebration for Mom's birthday at home. She had no suspicion that a big party was going to be held two days later. My daughter Kathy invited Mom to spend the following night with her, so I could get everything ready for the surprise party.

By the evening of the 10th, the house was decorated to the nines, the caterer was set to feed the guests, and the harpist I hired was ready to entertain. The guests were instructed to arrive thirty minutes before Kathy brought Mom home.

Bob Munger and his wife were the first to arrive. Shortly after them came Jane Withers. Her escort was John Buonomo. John was and still is a good friend of mine.

I met John, through Jane, in 1974. John lived in Las Vegas but visited friends in California quite often. Later, you will know why it's important to remember his name.

Most of the guests arrived before Kathy brought Mom back to the house. When we saw Kathy parking her car in front, everyone prepared for the big surprise. Mom stepped into the entry hall and looked into the living room. Her eyes enlarged and a look of complete surprise covered her face. I knew she hadn't suspected anything. She was overwhelmed and elated.

The house was gorgeous. I live in no mansion, but I am fortunate to have a very nice, richly furnished home that is quite exquisite. When it's decorated for a party, it's even more beautiful.

One hundred balloons were on the ceilings of the living room, family room and dining room. Long metallic streamers attached to each balloon hung within a few feet from the floor. Fresh flower arrangements were on the dining room table, on the baby grand piano in the living room, and on the coffee table in the family room. And of course, there was a special 80th birthday cake for Mom.

All of my friends had heard about Bob Munger, but few had met him prior to that evening. Naturally, I introduced him as the producer who was going to produce my movie. Bob had extended conversations with several of my friends and expressed how excited he was about making the film.

Several hours passed, and Tom still hadn't arrived. I felt a bit disappointed because I had looked forward to seeing him, and he had seemed so anxious to attend.

"What do you suppose happened to Tom," I asked Bob.

"I have no idea," he replied. "He told me he was going to be here."

I felt uneasy. I thought Tom would call and let me know he wasn't coming. But, I was delighted to see Mom enjoying the best birthday party she ever had.

The highlight of the evening was the harpist. He didn't play the typical golden harp. He mastered the Paraguayan Harp, a large wooden harp that was popular in Paraguay. It was attached to an amplifier, and the sound was beautiful. The harpist, Orlando Ortiz, could play anything, from show tunes to classics. He had appeared all over the world and had earned a gold album in South America. He held a captive audience that was entranced by his talent.

The evening was almost over when Tom arrived. I was happy to see him, but I was disappointed that he missed most of the festivities. I introduced him to my mother and to most of my friends. Then we went into the dining room so he could get a bite to eat.

Bob Munger and his wife at
my mother's birthday party at my home.
Taken August 9 1997

Before Tom picked up a plate, he handed me a copy of the first forty-six pages he had rewritten on the script. It was dated August 9, 1997. He had made some changes since Bob had faxed me the draft dated July 31. We had a brief discussion about his changes then I put the draft in a cabinet so I could read it later.

While we were in the dining room I introduced Tom to Jane Withers. Besides being a former major child star and "Josephine The Plumber" on the long running Comet commercials, Jane is an avid doll and teddy bear collector. Tom had written a speculative script about teddy bears that included scenes with Jane playing herself. He was happy to meet her in person. Thanks to that introduction, Jane gave Tom the card of a producer who was in search of a writer. Later, Tom secured the job.

That evening Tom wasn't as jovial as usual, and he seemed to be somewhat tense. I thought that perhaps he arrived late because of a personal problem. He was also eager to leave. I don't think he was at the party for an hour before he was out the door and on his way home. I don't recall that he even spoke to Bob Munger.

The party was a smashing success and everyone had a good time. I accomplished my goal. It was the best birthday party Mom ever had.

Jane took snapshots at the party and later filled two silver-framed photo albums with pictures. She gave one to my mother and the other one to me. That was one of my mother's favorite gifts. She had pictures of the party to show her friends in Illinois.

I also videotaped the party, and shortly after Mom returned home I sent her a copy. Every so often, I watch the video and relive that evening. My efforts to make my mother feel special and important were an overwhelming success, and for that I was delighted.

CHAPTER 57

Shortly after Mom's birthday party I received a call from Bob Munger. He was upset. He had been unsuccessfully trying to contact Tom Hubbell.

"I've called and left messages on his machine," Bob remarked. "I've emailed him several times and I even went to his house and left a note on his door. He has to know I'm trying to get in touch with him. I don't understand why he won't call me."

I thought Tom was probably busy with other projects, and Bob would hear from him soon. But that didn't happen. Bob's efforts to contact Tom were futile.

"I don't understand what the problem is," Bob said. "I don't know of anything I did that could have offended him. For some reason he's avoiding me. I think it's time I found another writer to finish the script."

I wanted to call Tom but I didn't have his phone number. I still wondered why Bob didn't want Tom and me to have personal contact or work together on the script.

I felt very uneasy about changing writers in the middle of the stream. I was pleased with the work Tom had done. He had kept my original story intact, and he was resolved to write the script to my satisfaction. I had already given Bob fourteen thousand dollars to pay Tom. I couldn't understand why he would walk out in the middle of the script. I was puzzled by the entire situation.

Within days Bob called to tell me he had found another writer. His name was Jim Hardiman. Bob told me he had already given the script to Jim so the rewrite could be completed. Jim lived near Palm Springs. Fortunately, my home in Ontario was the halfway point between Bob and Jim.

"Jim's an excellent writer," Bob said. "He's written several published books, and he wrote a screenplay that's been made into a movie. I gave him everything and I told him everything you want the script to be. He will do a good job for you."

I felt Bob was truly concerned about my mission and, as far as I knew, it wasn't his fault Tom had disappeared.

"I appreciate your concern, Bob," I replied. "I trust you. I know it won't happen but if your funding falls through and I get funding of my own, I'll see that you're still a part of this movie."

I felt secure with Bob's judgment and I didn't question Jim's writing ability, but I was concerned that I hadn't met him. Just as with Tom, Bob didn't want me to contact Jim. There weren't going to be any meetings, and I wasn't able to approve anything he was going to write.

For the next few weeks I waited with apprehension to see what was going to happen to my script. Finally, I received a call from Bob.

"Jim has the script almost finished," he said. "I think you're really going to like what he's done with it. Now you've got a good commercial script."

I breathed a sigh of relief. I was finally going to have the finished product I had agreed to pay for. I hoped Bob was right

On October 8, 1997, Bob drove from Westlake to my house in Ontario. He spent the night. Early the next morning I drove the two of us to Cathedral City. During the drive Bob told me he was still puzzled that Tom had never returned any of his messages.

"Tom's had a lot of personal problems," Bob said. "I did everything I could do to contact him, and I know he's purposely avoiding me. I think he is doing drugs."

I had no reason to disbelieve Bob. He had known Tom for over 20 years. I had only known Tom for a few months. But I still liked the work Tom had done on my script, and I felt his disappearance was a loss.

About an hour and fifteen minutes after we left my house, we arrived at Jim's home. It's located in a very nice gated complex in Cathedral City.

Bob made the introductions and Jim invited us into his home. We went into the living room. Bob and I sat on a sofa and Jim sat in a chair that faced us.

After a short conversation, I learned that Jim was the author of 13 published books. MGM Studios bought his first novel, titled "O-Bake," the Japanese word for ghost. In 1983, it was made into the movie "The House Where Evil Dwells," starring Edward Albert,

Susan George and Doug McClure. The Academy of Science Fiction and Fantasy Films awarded it "The Golden Scroll of Merit." Another book of Jim's "Song of Three Shepherds," about The Miracle of Fatima, is now on it's way to becoming a film.

I learned a lot about Jim's character from that conversation. He wasn't just an accomplished writer; he also was a good spiritual person. I felt I could trust him and I wasn't as concerned that Tom hadn't completed the rewrite on my script. I was anxious to see the fruit of Jim's efforts.

Bob handed me a copy of Jim's draft. I sat on the sofa next to Bob and read the script while they continued their conversation.

The moment I read the first page of the script, I was disappointed. The beginning had been changed. The film no longer started at the Vatican. It now began in Nazi Germany with troops marching through Berlin. The more I read, the more uneasiness I felt. I didn't say anything. I kept reading, hoping to find something that resembled my story. That didn't happen. By the time I read the last few pages my stomach was in turmoil.

My story was gone. Even the first 46 pages that Tom had written had been changed. I was so upset I almost broke out in a sweat. I laid the script on the sofa beside me and sat in silence. I didn't know what to do. My story had been destroyed.

Bob and Jim knew something was wrong because I didn't say anything. Finally, Bob broke the silence.

"Well, what do you think of it," he asked. "Now we have a good commercial script that's marketable."

I didn't know what to say. Jim had spent a lot of time working on that draft, and I didn't want to offend him. But I was devastated, and I had to speak my mind.

"If someone wants to make a horror film, it's a good story," I declared. "But it has nothing to do with my script or my treatment."

"What do you mean," Jim asked. "What script and treatment?"

"The script and treatment Bob gave you," I replied. "This is supposed to be a polish of my script, not a totally new story."

I could tell Jim was baffled by my comments. He quickly glanced at Bob then looked back at me.

"Bob didn't give me a script or treatment," Jim said. "I didn't know they existed. He just gave me the half of the script that Tom wrote and told me to finish it."

I proceeded to tell Jim about the mission that I was trying to fulfill. I told him every strange thing that had happened to me, from my near-death experience up to and including that moment.

After Jim heard my story he told me about some of his own supernatural experiences. I could tell that Jim believed I had told him the truth.

"Bob, why didn't you give me George's treatment and script?" he asked. "You didn't even mention them to me. I agree with George. This script doesn't serve his purpose."

"I thought I gave you everything," Bob sheepishly uttered.

I felt something was fishy. I didn't understand how he could not have known he didn't give Jim my material.

"A script that doesn't tell my story is worthless to me," I responded.

"I wasn't told this was supposed to be a polish and rewrite of someone else's script," Jim said to Bob. "I knew nothing about it. All you said to me was that it was supposed to be about a girl who was being pursued by the Devil. I had the impression that I had carte blanche to write anything."

I knew Jim was telling the truth. I don't know how Bob Munger sat on that couch and faced either one of us.

"Bob, why didn't you give Jim my script and my treatment," I asked. "You told me you gave all my material to Jim when you talked to him about finishing my script. How was he supposed to write the script I was paying for if you didn't tell him about it?"

Bob was noticeably embarrassed and uncomfortable and I had the feeling that his forgetting to give Jim my material was no accident.

"I agree with George," Jim said to Bob. "The only way this script can be done is to keep George's story intact. All of his experiences have to remain a part of his story. Without them, the messages people are supposed to get from seeing his movie are lost."

Jim offered to take the time to redo the script, but he couldn't do it without knowledge of what he was to write. Bob was between a rock and a hard place. He knew he had screwed up. He was caught and he had to do something about it.

"I'll send you copies of George's material," Bob replied

Jim wasn't happy that he had wasted so much of his time. He had every right to be angry. It wasn't his fault my story had been destroyed. Bob was responsible for that.

On the way home I didn't say much to Bob. I wasn't happy with him, but I still wanted to give him the benefit of the doubt. I didn't want to believe he had purposely tried to destroy my story. If that were so, he was doing dirty work for the devil.

"What was I supposed to do with the script you had Jim write," I finally managed to ask. "It had nothing to do with my story."

"Don't worry about it," he replied. "It's only the first draft. Things can always be changed. We'll work on it a lot more before we make the movie."

I was still worried about it. If he didn't keep my story intact, my money was wasted. I still feared that if I made him angry he would walk out on his commitment to make the film. So once again, I bit my tongue and kept my mouth shut.

We arrived at my home to find John Buonomo awaiting my return. He was on his way to Los Angeles from Las Vegas and had decided to stop at my house for a brief visit. I quickly reintroduced John and Bob to one another. They had met before at my mother's birthday.

I prepared us a drink then we sat in my living room and talked. Within minutes Bob stated he had to leave. I walked into the entry hall and started to open the door but he raised his hand and whispered.

"Before I go, give me a check for six-thousand dollars. I have to pay Jim for his work."

I couldn't believe that all Bob cared about was getting more money. He showed no concern the script had to be completely rewritten. I hid my anger, but there was no way I was going to give him a check for the full balance due on the contract.

"What happened to the fourteen thousand dollars I gave you to pay Tom?" I asked.

"I already gave him all of it," Bob replied.

"I don't want to pay the full balance today," I told Bob. "I'll give you two-thousand dollars now and the rest when Jim fixes the script. I haven't received anything that's even close to what I've been paying for."

Bob was reluctant to take a partial payment but because he knew I wasn't happy with the situation, he agreed to accept it. As soon as I wrote the check, he was out the door and on his way home.

I returned to the living room and resumed my conversation with John. I was so stressed I couldn't sit still. I paced back and forth and spilled my guts to him. Within minutes he knew everything that had occurred in Cathedral City.

"My story is destroyed," I kept repeating. "Bob has taken my money and destroyed my story."

John had read my original script and he knew my story well. He also knew everything that had happened to me before and after I started to work on it. He was disappointed, but he wasn't surprised that I was encountering another problem.

"When I met Bob at your mother's birthday party, I wasn't impressed," John said. "There was something about him that I didn't like."

I thought it was interesting that John felt that way. He wasn't the first person to tell me that. Shortly after Mom's party, Bob and Cindy Mills warned me about Bob.

"I don't know what it is, but there's something about him that isn't good," Cindy had said. "I wouldn't trust him if I were you."

Several other friends told me they weren't comfortable around Bob and didn't trust him at all.

I still gave Bob the benefit of the doubt. I didn't understand how he couldn't have known he hadn't given Jim my material. But I didn't think he did it intentionally. What would he have to gain? Why would he lie? Why would he deliberately try to destroy my story?

Bob knew that, through my story, I wanted to change people's lives and bring them closer to God. If he purposely tried to destroy my story, he had lied to Jim and me, and he was helping the devil. I seriously questioned Bob's motives, but I didn't think he would want that on his conscience.

John could see I was a nervous wreck. He did everything he could to calm me down. Nothing helped. I kept pacing the floors trying to believe I was going to wake up and find that it was all a bad dream.

Jim had given me his phone number so I decided to call him. I desperately needed his support. I placed the call at 1:05 p.m. on October 9th.

When Jim answered my call, I told him how upset I was with Bob, and how concerned I was about my script.

"I hate to bother you," I added, "but I'm having a hard time dealing with the problems that Bob has caused."

"You don't have to apologize for anything," Jim replied. "I agree with you; you have reason to be upset. What he did wasn't right, but hang in there. I'll fix it."

"Is it ok if I bring you a copy of my script and my treatment instead of waiting for Bob to mail you one?" I asked. "I want to make sure you receive it."

"Sure," Jim replied. "I have no problem with that."

Jim was very understanding and supportive and, by the time I hung up the phone, I felt much better. I called him four more times that evening. It's a wonder he didn't tire of me, but every time I spoke with him he refueled a spark of hope that everything would be ok.

That afternoon John read Jim's draft of the script. He agreed that it was a well-written horror tale based on my idea. But, he also agreed that my story was gone. And, because of Bob, the time Jim had spent on my script had been wasted.

John, as well as everyone else, didn't believe Bob accidentally forgot to give Jim my material. They felt it was intentional. It was possible they were right, but I still wanted to believe Bob was being honest with me.

CHAPTER 58

The following day, John accompanied me on the trip back to Cathedral City. He had a friend who lived only a few blocks from Jim's house. I dropped him off to visit with his friend and I took my script and treatment to Jim.

When I arrived at Jim's home I gave him the copies of my script and treatment. I also gave him a condensed copy of my Biography.

I wasn't aware these even existed," Jim said. "If Bob had given them to me, the script I wrote would have been a lot different."

I stayed for about two hours and visited with Jim. He was very supportive. Talking with him made me feel much better. He apologized for what had happened to my script but he was as much a victim of the situation as I was. He had done the job he had been told to do. Yet, he offered to rewrite the script so my story could be saved.

By the time John and I returned to Ontario, I felt more at ease. I had confidence in Jim but I was still upset with Bob Munger. I was curious to know the real reason Tom Hubbell had abandoned Bob.

That evening, I called Bob and told him I had personally taken copies of my material to Jim. He instantly retaliated in a rage.

"Who's supposed to be the producer of this movie? You or me," he yelled. "If you're going to take things into your own hands and go around me, I won't have any more to do with this movie."

I was shocked by his reaction. I hadn't thought he would care if I took my script to Jim.

"I didn't intend to create a problem," I replied. "I thought I would save you the trouble of mailing everything."

Instead of telling Bob to go to hell, like I should have, I apologized. He still held the threat of canceling the movie over my head. Besides that, I had already paid him most of the money for the rewrite. I had a lot to lose and he knew it. I had lived up to my part of our agreement but most of his promises had been verbal, and he hadn't lived up to any of them. His tracks were covered, and I was at his mercy.

CHAPTER 59

That night I was too upset to sleep. All I did was pace the floors and pray. I had already given Bob sixteen thousand dollars to pay writers for the rewrite. If he walked out on me, there was no way I was going to get my money back. He had promised to produce the movie, and had assured me the script was going to be based on my story and written to my satisfaction. I no longer believed Bob and I was devastated.

The next morning when John came downstairs, I was sitting in the living room. I had been up all night, too upset to sleep. John did his best to assure me that everything would be ok.

Once again Tom Hubbell came to mind. I wondered why Bob didn't want Tom and me to work together, and why Tom had abandoned Bob.

"I wish I had some way to get in touch with Tom," I said to John. "Bob didn't want us working together, and he didn't want us to have each other's phone number. Tom was doing a good job on my script, and he had no reason to walk out on me. Maybe Bob pulled a dirty deal on him."

"Wait a minute," John said. "I may have Tom's phone number. When he was here for your mother's birthday party, I think he gave me one of his cards. If he did, I should still have it."

John went upstairs to his room and returned a few minutes later holding a business card in his hand. My heart pounded with anticipation.

"I found it." John said, as he handed me the card.

Within minutes I was on the telephone with the first call I ever made to Tom Hubbell.

Tom didn't answer the phone. I left a message on his recorder and asked him to return my call. He called me back about 90 minutes later.

"I can't believe you called me," Tom said. "I wanted to call you long ago, but I didn't have your phone number until a couple of days ago. I found it written on the inside of one of your scripts and I was going to call you."

"I've wanted to call you long before this," I replied. "But I didn't have your number either." I then explained to Tom how I got his number from John. Later, Tom told me that he had purposely given his card to John, hoping I would get access to his phone number.

"It's really strange that you called me. I was going to call you today. I have never stopped working on your script. I've been doing some research and I've found things that back up a lot of the things you wrote about. I have a record book that's five inches thick and it's filled with research I did for you. I didn't call sooner because I wanted to have everything together first."

I was relieved to hear that Tom still had an interest in my mission. I filled him in on everything that had transpired since I had last seen him, including the fiasco Bob had caused. He didn't seem to be surprised. That made me even more curious to find out why he was avoiding Bob.

"What happened to you?" I asked. "Bob tried to get in touch with you for several weeks and you didn't respond. I thought you didn't want to work on the script anymore."

"I've never had your script out of my mind," Tom said. "But I had to get away from Bob. I don't ever want to see him again."

Now I was really wondering what the problem was.

"What happened?" I asked. "What did he do?"

"It had nothing to do with you and your script," Tom replied. "It's a personal thing between Bob and me."

Tom didn't want to tell me what the problem was, so I didn't ask. We were finally in touch with one another, and that was all that mattered to me.

"I want to get together with you so you can see all the research I've done," Tom said.

"Are you doing anything this evening?" I asked.

"No," he answered.

"I'm free this evening," I said. "Why don't you come out tonight?"

Tom drove the 75 miles from Woodland Hills to my house that evening. When he arrived, he gave me the notebook filled with the research he had done. I was impressed. I could see that he had spent lots of time on it.

"I really believe in what you are trying to do," Tom said. "I want to help finish the rewrite on you script."

"I would love to have you help finish it," I replied. "But all of my money is gone. I have nothing left to pay you with."

"I don't want to help you for money," Tom said. "I want to help finish it because I believe in you and your mission."

"That makes me feel very good," I told Tom. "But I couldn't let you work for nothing."

"It doesn't matter," he replied. "I've seen how dedicated you are to this project. You've gone through hell and you keep on going. You've never given up and quit. You have a mission and you're determined to fulfill it. Now, it's my mission to help you. I don't want any pay. Just knowing I helped you is worth more to me than any amount of money."

CHAPTER 60

Soon, Tom and I were working together on the rewrite. We started from page one. All the drafts of the script Tom had written to date were stored on his computer. Therefore, we worked on the script at his home in Woodland Hills.

We had barely started working together when Bob called to tell me Jim had fixed my script. It was November 7, 1997. Bob asked me to drive to Westlake to give him the final four thousand dollars so he could pay Jim.

Instead of holding that meeting at the Westlake Hyatt Hotel we had lunch on the patio of a restaurant by the lake. I don't know why we didn't meet there before. The atmosphere was great. We sat at a table near the water, and I enjoyed hand feeding the birds as much as I enjoyed eating my food.

Bob seemed to be in a much better mood. He didn't mention any of the problems we had encountered. He probably was afraid we would end up having a confrontation.

According to Bob, he was an expert in marketing and advertising. Instead of reviewing my script he discussed ways to improve the advertising for my tuxedo business.

"If you had to pay for my expertise you couldn't afford me," he said. "I'm giving you thousands of dollars worth of information for nothing."

If the information he gave me was worth thousands of dollars, I'm in the wrong business. He didn't suggest anything I hadn't already tried. I'm sure he rationalized that he had given me thousands of dollars of information that would compensate me for my twenty thousand dollars he wasted. If that was his intention, it didn't work.

"Jim did a good job on your script," he remarked. "Now you have your money's worth. Give me a check for the four thousand dollars and I'll pay him. You're lucky to have someone of his caliber write a script for six thousand dollars."

"But what if I don't like it," I asked.

"If there's anything that needs to be rewritten we can do that later," he replied. "We'll probably change lots of things before the movie's finished."

I couldn't believe Bob didn't give me a chance to read the script before he asked me for the money. I knew Jim had spent a lot of time working on that script and he deserved to be paid, so I reluctantly wrote the check and handed it to Bob. The ink didn't have time to dry before he was out of his seat and on his way home.

That evening I read Jim's new draft. It was basically the same story I had read before, with more of my material added to it. It was still a horror story, and not the spiritual story that I wanted.

Jim is an excellent writer, and he could have improved my script beyond my wildest expectations if he had received my material and the right instructions in the beginning. He wasn't responsible for the damage that had been done to my story and I couldn't ask him to spend any more time on it. The only hope I had of securing a satisfactory script was to pursue working on a complete rewrite with Tom.

I made two or more trips a week to Tom's house. Sometimes we worked until 2 and 3 a.m. Many nights I drove home exhausted, but I was always pleased with our progress. I felt fortunate to have Tom collaborating with me. His ideas and efforts were making the script everything I had wanted it to be.

I didn't want to do anything behind Bob's back, so I made a phone call and told him I had made contact with Tom. He was overwhelmed. But he didn't make one negative comment. He didn't even ask if I knew why Tom refused to return his calls. I didn't want problems, so I didn't mention it either.

"How did you get in touch with Tom?" Bob asked.

"He gave his card to a friend of mine at my mother's birthday party," I replied. "He had never stopped working on my script, and he volunteered to work with me on a complete rewrite for nothing."

"I don't see anything wrong with the script Jim wrote," Bob remarked.

I avoided the reasons I wasn't happy with Jim's script. I didn't want a confrontation with Bob.

"Jim did write a good script," I replied. "But it isn't the story I want to tell. I don't want to make a horror film. I want a movie based on truth that'll change people's lives. Tom and I work well together, and the script is going to be very good."

"Well, that's good," Bob said. "When you're finished with it, bring it to me. Maybe we'll end up with a good script and we can still get this movie going."

After hearing those statements I was glad I hadn't said anything to offend him. If he lived up to his commitment to produce the film, I didn't care if I lost my money and had to see that the script was rewritten on my own. I was glad Bob was still interested in making the movie.

During our next work session I told Tom about my conversation with Bob.

"I want to be up front with Bob," I said. "I have to be honest with him. Besides, we have no reason to hide the fact that we're working together."

"It's best to not have any problems," Tom replied. "You want your movie made and Bob has the connections to get the job done. When we're finished, give him a copy of the script. If he's telling the truth and he helps get your movie made, that's all that matters."

CHAPTER 61

During the months Tom and I worked together I kept Bob updated on our progress. He always assured me he still had every intention of making the movie.

No matter what I said to Tom, he still didn't want anything to do with Bob. All that mattered to him was that Bob kept his word. He still insisted the problem between them was a personal one and it had nothing to do with me.

I was exhilarated with the progress Tom and I were making on the script. It was a thriller, but it was spiritual, it was believable, and it based on fact. I knew it was going to make a darn good movie.

Financially, things were not doing as well. I had more losses on the apartments and I had lost twenty thousand dollars to Bob Munger. I had no money left. That was about the time I received a call from Nick Cassa.

Nick and I first met at the Wilshire Country Club in the fall of 1996. He is a Class "A" PGA golf professional, and he was trying to organize a celebrity golf tournament for a charity fund-raiser I was working on. The tournament never became a reality, but Nick and I became good friends.

When Nick learned about my mission, he became very supportive. He knew people would benefit if my efforts were successful.

I gave Nick a copy of my script. Shortly after he read it he sent a copy to a group of investors headquartered in Las Vegas. They had access to millions of dollars and he thought they might fund my movie. To my surprise, one of the investors called to tell me they were interested. I was told the money would definitely be funded to make the film. I was thrilled, but my expectations were short lived. After weeks of waiting and a slew of false promises, nothing happened.

Nick was upset that his friends had fed me a line of BS, but he had done his best to help me. It wasn't his fault. I was disappointed, but I added it to the list of false promises people had made to me in the past.

After the Las Vegas investors flaked out on us, Nick wanted me to meet his good friend, Kermit Alexander. Kermit is a former pro-football star who had played with the Rams, Eagles and 49ers. Nick had already mentioned my project to Kermit and he had expressed an interest in it. Kermit was also involved with a group of investors.

I met Kermit for the first time in March 1998. He needed a tuxedo for a formal function, so I met him at my Chino Store. After his fitting, we went to a near-by restaurant. He wanted to hear all about my project so I told him the entire story, from beginning to end. He was mesmerized.

I had barely finished telling Kermit my story when he told me he wanted to get the funding to produce the film. Once more someone wanted to help me. This time I had high hopes he would succeed.

CHAPTER 62

The following week Tom and I finished the script. I gave a copy to Kermit and within a couple of days I received his response. He loved it. He was especially touched by one thing. In my story, gangbangers murdered Amelia, the mother of my main character, Amy. In real life, gangbangers had murdered Kermit's mother, sister, and two nephews.

In March or early April of 1998, I went to a banquet at the Beverly Hills Friars Club. At that banquet I met Henry Silva, a well-known character actor who has appeared in over 80 films. Among his credits are "Taras Bulba," "The Manchurian Candidate," and "Above The Law," with Steven Segal. I had a copy of my script with me. A friend of mine knew Henry and asked him if he would read it. To my astonishment, Henry took the script home with him that night.

I thought it would be at least a few weeks before Henry read the script, but only a few days later I received a message from him on my recorder.

"George," he said. "This is Henry Silva. "I read your script "The Devil's Reign," and I have to talk to you about it. This is one of the best scripts I've ever read. I've read hundreds of scripts and this is one of the few that I instantly saw as a blockbuster. This is the kind of movie that makes stars. This is something else. This movie has to be made! Please call me."

When I heard that message, I was on cloud nine. The instant the recorder shut off I returned Henry's call. His continued praise of my script made me feel even better.

"I don't usually read a script unless it's funded and I'm reading for a part," he said. "But for some reason I agreed to read yours. I even set aside enough time to read it from beginning to end. From the minute I started reading it, I was hooked. I couldn't put it down. I'm telling you this is a blockbuster. This movie has to be made."

Henry wanted to play the role of Fr. DiMarco. In the story Fr. DiMarco is Amy's uncle, a troubled alcoholic priest who goes through life-changing experiences after he comes to Amy's aid. Since that conversation, Henry and I have become friends.

I met Bob Munger at the Friars Club that same month. We had lunch and I gave him four copies of the finished script that he had requested. I didn't tell him Kermit was trying to secure funding for the production. I didn't want him to proceed with the production because of the prospect of getting money through Kermit. Bob took the copies of the script and told me that he would contact me as soon as he read it.

The following week, Kermit met Shirley, Tom, and me at the Friars Club for lunch. He told us his investors had made a commitment to fund the production of the movie.

I was so thrilled I had to pinch myself to make sure I wasn't dreaming. I sat at that table and thought of all the things I had been through to get to that point. Finally, something good was happening to me. I was so happy, I cried. I put my hands over my face and tried to hide my emotions, but I was so overwhelmed it was impossible.

"George has every right to let it all loose," Tom said to Shirley and Kermit. "He has been through a lot and he has stuck with his story. He deserves to have something good happen to him."

I contained my emotions then I called my mother in Illinois to tell her the good news. She was almost as excited as I was.

As I was leaving the Friars Club after that meeting, the hostess of the club called to me.

"George! Come here a minute."

I was curious to hear what Rose wanted.

"Who was that man you introduced me to the day you had your scripts here," she asked.

"Do you mean the man with the dark hair that I gave the scripts to?"

"Yes, he's the one," she answered.

"That's Bob Munger. He helped produce "The Omen," and he's the producer of my movie. Why do you ask?"

"I don't know," Rose said. "There was something about him that didn't feel right. I felt bad, almost evil vibes coming from him. Please be careful with him."

That wasn't the first time someone said something like that about Bob Munger. It reminded me of the comments my friends made after my mother's birthday party.

Perhaps it's time I started listening to people's warnings, I thought. *Maybe Bob didn't "accidentally" forget to give Jim my material. Maybe he's working for the Devil and he's trying to destroy my story and me.*

Bob had made every effort to convince me that he was my best friend and was looking out for my best interests. I had listened to him because I wanted to believe he was telling the truth. I wanted to get my movie made. But my gut feeling had always told me that something was wrong. If I had listened to my heart instead of my mind, I never would have trusted him.

Then, I thought perhaps we were all wrong. Bob was supposed to be a devout Christian. Maybe the Devil was making him look bad so I wouldn't trust him. If he knew I suspected he was evil, he would never produce my film.

The following week, Bob called me. He wanted me to meet with him to discuss the script.

I no longer trusted Bob enough to go to that meeting alone. I took Shirley with me. When Bob arrived we had lunch and he started his critique of my script.

I knew it was a good script, and I was waiting to hear him praise the job Tom and I had done. To my amazement, he didn't say one nice thing. He tore the script apart from beginning to end.

I knew something wasn't right. He finished ripping the script apart then he looked me straight in the eye. I felt an evil coldness that came from deep inside him and I thought about the comment my friend at the Friars Club had made. I felt like I was sitting in the booth with the Devil.

"Well, George," he said arrogantly. "You're going to have to throw out all that Vatican stuff in the beginning. As a matter of fact, you're going to have to do another rewrite."

"I don't mind changing some things, Bob," I replied. "But I'm not changing anything, and I'm not doing another rewrite, until I have a contract from someone to do this film. You had me write a treatment for you. That wasn't good enough. Then you had me spend $20,000 for a complete rewrite. Now, you tell me it's not good enough and you want it rewritten again! I'm not changing anything until I have a contract from you. Rewrites can be done later."

That's when I saw the real Bob Munger. He looked as if fire was going to shoot from his eyes and his attitude was anything but godly.

"You change it or else," he coldly responded.

At that point I was fed up with Bob Munger. I felt sick to my stomach because I had kept my mouth shut and took his bullshit for over a year. I wasn't going to take it any longer.

"I'm not going to change anything," I retorted. "Not until I have a contract."

"Then you find someone else to make your movie," he replied.

With that, Bob stood up, mumbled a couple more sentences, and left. He showed no respect for Shirley or me. We were both amazed that he had been so cold and heartless. But we had seen the real Bob Munger.

"It felt like we just had a meeting with the Devil," I said to Shirley.

"It sure did," she replied.

I finally saw Bob for what he was. He proclaimed himself to be a devout "Born again Christian." But to me, he was the Devil's Disciple.

CHAPTER 63

A few days later I was at Tom's house making a few minor changes in the script. Before I left I told him everything about the meeting Shirley and I had with Bob.

"I couldn't believe Bob could be so cold," I said to Tom. "I gave him $20,000 to pay writers for a rewrite and I trusted him. I don't understand how he could do me so dirty and feel no remorse. He can't have a conscience. I think he purposely ripped the script apart so he'd have an excuse to walk out on his commitment."

"How much money did you give him," Tom asked.

"I gave him $20,000. You got $14,000, and I gave him $6,000 more to pay Jim."

"Wait a minute," Tom retorted. "I didn't get anything near $14,000, and he never offered to pay me $20,000 in the first place."

"I gave him $14,000 and he told me he gave all of it to you before you stopped having contact with him," I answered.

"Well," Tom responded, "It looks like Bob ripped off both of us. Now that I know he isn't going to keep his word and help you get your movie made, I think it's time you knew the truth of what happened between Bob and me. I didn't want to say anything to you before, because he was still telling you he was going to get your movie made. I didn't want to say anything that would jeopardize that. Now, it doesn't make any difference."

I didn't expect to hear anything too astounding. After the experiences I had with Bob, I thought he was capable of doing anything. But I didn't expect to hear what Tom was going to say.

"When Bob first called and asked me to take this job, he told me you had a script that you wanted rewritten, that you were rich, and money was no object. He told me I basically had carte blanche to write anything I wanted.

When I first took the job, you and I hadn't met, and I didn't know anything about you. All Bob wanted was for me to throw out pages. He didn't care if they were good or not. When I had the first meeting with you and Bob, and I heard how dedicated you were to your mission, I felt like hell. You had your heart and soul wrapped up in that story and I realized how much it meant to you. I saw there was a lot more to you than Bob had told me.

I also found out that, as far as Bob knew, you had borrowed the money to pay for the rewrite. You had to pay that money back, and Bob didn't care.

I was impressed even more when I found out all the things that had happened to you. You were going through hell and you still didn't give up. You were determined to complete your mission.

Every time we had a meeting you told me every detail of an experience that a scene was based on, and why it had to remain in the story. I wanted to take the time to do a good job rewriting your script. But after every meeting, as soon as you left, Bob told me not to listen to you. He told me to throw out almost everything you wanted to stay in the script. He said he could handle you through Shirley.

After every meeting, I went home feeling sick to my stomach." Tom continued. "I knew Bob was taking your money, and he didn't care if he destroyed your script. Every day he was calling and hounding me for more pages.

I felt bad about what Bob was doing, and I knew I had to do something about it. I had two ways to go and I had to make a decision: I could keep working with Bob, give him his pages and let him continue his wrongdoing, or I could get away from him. My conscience wouldn't let me be a part of his dirty work, so I had to get away from him.

We had been friends for 25 years, but I no longer respected him. I knew I could never be around him again. I didn't have your phone number, so I had no way to call you and make other arrangements to work on the script."

"By the way," I said to Tom, "Bob told me he thought you disappeared because of a drug problem."

"No way," Tom responded, "I've never had anything to do with drugs."

I was shocked to learn that the whole time I had known Bob, he didn't care about my story, my mission, or me. All the time he had been telling me he would do nothing but look out for my best interest, he had been deceiving me, ripping me off, and trying to destroy my story.

187

I was glad I hadn't told Bob that Kermit had a commitment for the funding to produce the film. If he had known there was a possibility of big money coming, he might not have walked out on his commitment and I would have remained his stooge.

Bob Munger is definitely one of the most evil persons I have ever known. But if it hadn't have been for him, I would never have met Tom Hubbell and Jim Hardiman. They are two of the best people I have ever met. I will forever be indebted to Tom Hubbell.

Bob never realized that when I was able to come up with the $20,000 to pay for the rewrite, he was put to a test -- a test that he failed.

As I said earlier, when I didn't think I was going to be able to pay the $20,000, his comment to me was, "You have had all kinds of things happen to you that make you believe in what you are doing. I haven't had any of those things happen to me. If God wants this movie made, he will see that you get the money. If that happens, it will be a sign to me that I am to make this movie."

I never did receive any of the money from my Hollywood friend. Bob never knew it, but I had been able to pay the $20,000 on my own.

Bob had his sign, and God tested him. But it was a test Bob failed. Instead of living up to his own words and seeing that the movie was made, his selfishness and greed made him help the Devil destroy it. Bob took my money and destroyed my script, but he lost something no amount of money can buy.

A year later, Mike Rappaport wrote a story about me in the The Daily Bulletin newspaper. In Mike's interview, Munger insisted he did nothing illegal and said he was only trying to help me. "I am not in the scam business," Munger said. "I told George what he had was no good."

If that was the case, why did he take my $20,000 to polish my script and promise to produce the film?

CHAPTER 64

I would have been more upset over what Bob did, but I still had a good script, and Kermit's investors were going to fund the production of the film. My movie was going to be made.

A few days later, I spoke to Jane Withers. I told her that I had the possibility of getting funding for the film. She suggested I speak to an experienced producer who had been highly recommended to her. His name was John Purdy.

I called John. The following week, Kermit and I had a meeting with him at the Friars Club. John found our project to be interesting and gave us tons of free advice. He took a copy of my script home with him. Two weeks later he sent me a complete breakdown of the projected budget to make the movie.

About that same time, I attended a charity function in Hollywood. I started a conversation with a man who was sitting at my table. He had fairly short hair and appeared to be in his thirties. His shoulders and arms were firm and strong. I thought that perhaps he worked out at a gym.

"What do you do," he asked.

"I wrote a script that's going into production," I replied.

"Oh? What's it about?"

The money was being funded to make my film and it was going into production, so I never thought any harm could come from telling him about my story. I told him everything, from beginning to end.

"That's a very interesting story," he said. "I'm writing a script that's very similar."

At that moment, I felt telling him my story was a big mistake. But it was too late.

"What do you do for a living," I asked.

"I'm a screenwriter," he responded.

"What's your name?"

"Andrew Marlowe," he replied.

"Have you ever had a script produced," I asked.

"Yes, "Air Force One.""

I didn't know what to believe. I had never heard of him. For a moment I thought I had met another Hollywood phony. Then I realized I had no reason to doubt his word.

Suddenly a strange thought passed through my mind. I wondered if he was working on a script with Bob Munger. If so, that would explain why Bob had been so cold and why he ripped my script apart. He needed an excuse to walk out because he was stealing my idea and working on his own story.

I wish now that I had asked Andrew Marlowe for an answer to that question. I couldn't believe I had met, by chance, a writer who was working with Bob Munger. So, I didn't mention it.

As I drove home from Hollywood, I kept thinking about that strange encounter. The more I thought about it, the more it bothered me. I called my friend, Cindy, from my cell phone and told her what had happened.

"If he really did write Air Force One," I added, "I probably don't have anything to worry about. I don't think a writer who is that successful would steal my story, especially since he knows my film is going into production."

Cindy agreed, so I felt better. But, I discussed that meeting with several other people before I stopped worrying about it.

CHAPTER 65

By the end of the month, I was working full time in the stores. With high school proms and many weddings, the tuxedo stores and the bridal shop were popping. I had very little free time.

The last week of June I received a call from Dea Martin, the hostess of the television show I appeared on in Raleigh, North Carolina.

"Hi, George," she said. "I'm visiting in Los Angeles with my son and I would like to see you before I go back to North Carolina. Do you have any free time?"

"What's your schedule like this week," I asked.

"We have no plans for the next few days," she replied.

"I have no plans either," I responded. "Why don't you come to my house on Thursday and we'll have a barbecue?"

"Oh, that sounds wonderful," Dea answered. "I have one favor to ask. I'm staying with a friend while I'm in Hollywood. Her name is Christina Miller. She's an actress and she's a lot of fun. I think you would really like her. Is it ok if I bring her with us?"

"Of course, it's ok," I replied. "The more the merrier."

I invited several other friends to the barbecue. They included Wendy Moss, the daughter of Jane Withers.

By the time Dea, her son Dakota, and Christina arrived, there were eight of us ready for a good time. After introductions were made, we all went into the kitchen so we could talk while the food was being prepared.

The kitchen was a bit small for eight of us, but we managed. I was busy making the barbecue sauce and others were chopping vegetables for the salad, shucking corn, preparing the ribs, or keeping the kitchen clean. Christina sat in a chair near the kitchen door and observed.

"Do you mind telling Christina about your script, George?" Dea asked. "She would like to know what it's about."

"I don't mind at all," I replied. "I'm always ready to talk about it."

I stood in front of the stove and kept stirring the barbecue sauce, as I talked to Christina.

"My script is based on a true story," I said. "I had a lot of supernatural things happen to me."

I was ready to tell Christina about the morning I woke up and saw LaDonna floating over my bed. Before I said anything, Christina asked, "Were you raised in California?"

"No," I replied. "I was raised in Illinois."

"What town was you raised in?"

"I was raised in a small town near Springfield," I replied.

I purposely didn't mention Taylorville because I thought she had probably never heard of it.

"What small town," she asked.

"Taylorville," I replied.

"Did you ever know anyone there named LaDonna?"

I couldn't believe what she had just asked me. I stopped dead in my tracks and so did everyone else. Except for Dea's son and Christina, everyone knew I had seen a girl named LaDonna floating over my bed, and that I had called back to Illinois and found out she had died.

I stopped stirring the sauce and looked straight at Christina. My heart was pounding. I felt like I was dreaming.

"Wait a minute," I replied. "I was just getting ready to tell you about a LaDonna I knew, but the one I'm talking about died 22 years ago."

I didn't think Christina could possibly be asking about the same LaDonna I had gone to school with.

"The LaDonna I am talking about died real young," Christina replied. "What's the name of the husband of the LaDonna you knew?"

"Julio Monge," I answered.

Christina's eyes widened and a look of surprise covered her face. "That's the same LaDonna," she replied. "She was my cousin."

My heart skipped a beat. I couldn't believe what I had heard. Everyone was just as shocked as I was. For a moment we all remained silent and stared at one another. We were dumbfounded.

"Oh, my God," I finally uttered. "I'm covered with goose bumps from head to toe."

"So am I," everyone else said in unison.

"Look at my arms," my friend Wendy said.

She pushed up her sleeves and held one of her arms against mine. We were both covered with tiny bumps. Everyone started showing his or her goose bumps and we all started bewilderedly laughing. We weren't laughing because it was funny, but because we had all witnessed something that was hard to believe.

Christina was still sitting in her chair. She looked back and forth at each one of us with a puzzled expression. She couldn't understand what was happening. She didn't know about the experience I had with the apparition of LaDonna.

After we calmed down a bit, I told Christina that I had seen LaDonna floating above my bed, the same day she died.

"You've got to be kidding me," she exclaimed. "You can't be serious.

"No, I'm not kidding," I replied. "It really happened."

"Are you guys sure you didn't make all of this up and you're pulling one over on me," Christina queried.

"No, Christina," everyone said in unison. "It's true."

"I can prove we are telling you the truth," I responded. "I can show you a copy of my Bio. Everything that happened with LaDonna is written in it."

I got a copy of my Bio and showed it to Christina. After she read it, she was just as bewildered as the rest of us. Though she found it hard to believe, she knew we had told her the truth. (Christina's notarized letter of validation is included in the documentation section at the back of this book.)

We had such a good time we didn't want that day to end. No one had any commitments for the next few days, so I talked them into staying through the weekend.

Another strange thing happened Saturday, July fourth. I had videotapes of the movies, "The Miracle of Our Lady of Fatima" and "The Song of Bernadette." Both films were about apparitions of the Virgin Mary and documented miracles that were connected with the apparitions. We decided to watch both movies that afternoon.

The second film we watched was almost over when I had an inspiration.

"Oh, my God," I said. "I just realized something."

Everyone looked at me.

"I've watched these films at least a dozen times and I can't believe I never realized this before," I said. "In both of these movies, the people very seldom speak of Mary as, 'The Virgin Mary.' They always call her, 'The Lady.' I just realized that in Italian the words for "The Lady" are La Donna."

"Wow! That's amazing," Wendy exclaimed.

Needless to say, once again everyone in the room was covered from head to toe with goose bumps.

I couldn't believe that 22 years had passed since I saw the apparition of LaDonna and I never realized her name meant "The Lady," until that moment. I don't think the relationship between the two names is a coincidence.

I called Kermit and told him about the strange occurrences that took place that afternoon. That evening he came to my house to meet Christina and the rest of my friends.

After an extended conversation about the events of the day, we talked about the movie. The funding had been promised to Kermit in April. It was now the beginning of July and not one dollar had been funded.

"Are you sure the investors are going to come through with funds to make the movie?" I asked.

"It's a done deal," Kermit said. "It shouldn't be much longer until they fund the money."

That weekend was special to all of us. We knew the things we had witnessed were not coincidental. Besides spending an enjoyable four days together, we shared a spiritual experience we will never forget.

CHAPTER 66

Later that July, I decided to take time off work and attend my fortieth Taylorville High School Class Reunion. As usual, everyone had a ball. But for the first time, I realized we were aging. Our youth was trading places with gray hair and wrinkles. We were hanging onto our last sparkle of youth. But no matter how old our bodies were, our minds were ageless.

After the reunion, I spent a week in Taylorville visiting my family and friends. But I made sure most of my time was spent with my mother. She loved being with me, and I was thrilled to see her looking healthy and happy.

At the end of my prior trips to Illinois, saying goodbye to her and not knowing if I would ever see her again tore my heart out. But this time, heading for home was exciting. Mom was flying home with me to spend a month in California.

Shortly after we returned to California I invited Christina, Dea and Wendy to a dinner to meet my mother. At dinner, Dea mentioned that a man named Georgio Bongiovanni was going to speak at a seminar in Laughlin, Nevada. Georgio is one of very few recipients of stigmata, a phenomenon in which a person's head, hands, and feet bear the marks of the crown of thorns and the nails of the cross. They bleed the same as Christ's wounds bled during the Crucifixion.

Wendy and I were fascinated with the Bongiovanni story and wanted to go to the seminar in Laughlin.

On August 5th, Wendy drove Mom and me to Laughlin. When we arrived at the hotel the temperature was 115°F. The heat was unbearable. We felt like we melted walking the short distance from the car to the hotel entrance.

We were barely inside the hotel when Mom's legs started to cramp. My heart sank to the floor. I knew the circulation in her legs was bad and I didn't like seeing her suffering with pain. After a short stop to rest her legs, she was ready to proceed. But I knew Mom was trying to hide her discomfort from Wendy and me. She was afraid her problems would put a damper on our fun. That was the furthest thing from the truth. I had to do anything to help her. I couldn't have had a good time if she was uncomfortable.

I borrowed a wheelchair from the hotel and insisted that Mom let me be her feet. Once she realized that I loved having the chance to help her, she was fine. She sat in the wheelchair and had no more pain. I was filled with gratification. I was blessed to have the opportunity to comfort her.

That afternoon we went to Georgio's seminar. I don't know that his story has been validated, but we found him to be a very interesting person. He claims to have had a divine intervention that left him with Stigmata. Like me, he was on a mission to help mankind. Like me, he was warning people to get their lives in order with God. I hoped everything he said about his experiences was the truth.

The following morning we headed for home. Wendy drove the car, I sat in the passenger seat, and Mom reclined on the back seat and relaxed. She still didn't feel quite up to par, but she was talkative and jolly. And, she occasionally nudged me on the back of my head with her foot. Each time, I turned to see her give me a wink and a big loving smile.

It was late evening when we arrived home. Wendy and I wanted to stay up for a while, but Mom was exhausted.

"Is it ok if I go to bed?" Mom asked.

"Of course it is," I replied. "You don't have to ask my permission to go to bed. Get some rest and we'll talk tomorrow."

The next morning I was in the kitchen fixing myself a cup of coffee when I heard a loud crash within the house. I immediately ran into the family room where Wendy was sitting.

"Did you just hear a loud noise," I asked.

"No," Wendy replied. "I didn't hear anything."

Suddenly I heard screaming coming from upstairs. It was Mom. My heart sank to my feet. I knew something was terribly wrong. I ran into the entry hall and up the stairs as fast as my feet could carry me.

"Please God! Please God!" I prayed. "Please let Mom be ok!"

By the time I reached the upstairs hallway, Randy was in Mom's bedroom. When I entered Mom's room I was devastated. She was lying flat on her back on the floor.

"Help me! Help me," she pleaded. "Something is wrong. A part of me is missing. I'm not all here."

I knew she had suffered a heart attack or a stroke. I knelt beside her to comfort her, but there was nothing I could do. She had taken a hard fall. I knew she shouldn't be moved until professional help arrived.

"Mom, I can't believe this is happening," I said to her. "I've prayed so much for you. Try to stay calm. I'll get help as fast as I can."

The initial shock started to subside and reality set in. I knew Mom was in critical condition. She needed help immediately. I had to calm down and think rationally.

I grabbed the telephone and dialed 911. I was so upset I didn't think about pushing the paramedic button on the alarm system. If I had done that, help would have been sent immediately.

The moment the 911 dispatcher answered, I told her there was a medical emergency and I gave her my address.

"What's the problem," the dispatcher asked.

"I think my mother had a stroke," I shouted. "Please get an ambulance here as fast as possible!"

"What are her symptoms," she asked.

"It doesn't matter! She needs help now," I yelled.

The dispatcher kept asking unnecessary questions that were wasting time. She may have been doing her job, but I didn't care about anything but getting help for my mother.

"There's no time for that bullshit now," I finally yelled. "We don't have time to waste. Send help fast!"

I ended that call and immediately called my sister Jean, in Illinois. I was so distraught I could hardly tell her what had happened. I stuttered and stammered as I described the scenario that was taking place in my home. Then I asked Jean to call the rest of the family.

Needless to say, by the time that phone call ended, Jean was a basket case. I was upset, but I was with Mom. Jean was two thousand miles away. She could do nothing to help Mom. She could only imagine what was going on in California. At least I was dealing with reality. She could only deal with conjecture.

I had barely finished talking to Jean when the paramedics pulled up in front of the house. Within minutes they were upstairs at my mother's side. I prayed I would hear positive feedback from their examination. But that didn't happen.

"We have to take her to the hospital immediately," one of the paramedics said. "I'm pretty sure she's had a stroke."

I wanted to wake up and find it was all a bad dream. But that wasn't going to happen. I had to face reality. One of the most traumatic experiences of my life was taking place, and I had to deal with it.

Mom was trying to talk to us, but her speech was garbled and impossible to understand. I knew she was physically and mentally going through hell. She was suffering agonizing pain and she was terrified, but there was nothing I could do. I never felt so helpless in my life.

When the paramedics lifted Mom to place her on the stretcher she let out a blood-curdling scream. Her right arm had been hurt in the fall. Watching her suffer was almost more than I could bear.

I drove my car and prayed as I followed the ambulance to the hospital emergency room. Check-in took only a minute, then a doctor was at Mom's side. Shortly after completing his examination, he confirmed the suspicions that Mom had a stroke. He then left the room to have her admitted to the hospital.

Mom was very agitated, but she was well aware of what was going on around her.

"Mom, do you know who I am," I asked.

"Yes, I know, I know who you are," she stammered. "I can't think, I can't think of your name but…but you're one of my children. I know you're my child."

I was relieved that Mom was mentally alert, but I knew she was a long way from being well. I prayed like I have never prayed before for her total recovery.

After tests, we learned that Mom was paralyzed on the right side. And x-rays revealed that, as a result of the fall, her upper right arm and shoulder were broken in three places. Because of her acute condition, the doctor opted not to put her to sleep to perform surgery and reset the fractures. He feared she would not survive the operation. Unfortunately, every time she moved she was in agonizing pain.

CHAPTER 67

For the next few days, Wendy and I stayed at the hospital and helped comfort Mom. Before the week was over, my sisters, Jean and Mary, flew in from Illinois. We spent most of our time at Mom's bedside.

After two weeks of acute care, Mom's condition improved and she was transferred to Casa Colina Rehabilitation Hospital in Pomona. Her doctor made no promises of a complete recovery, but we continued to pray and hoped for the best.

One day my daughter Kathy, Mary, Jean, and I were all at Mom's bedside. She suddenly looked at us with confident expectation.

"I'll be glad when I get better so I can fart in a sock," she said loud and clear.

She immediately knew the words she said were not the words she intended to say. She was embarrassed and attempted to apologize. But we were thrilled to hear her say a complete sentence. We didn't care what she said. She was starting to respond. Actually, we thought her remark was quite funny and we laughed hysterically.

Mom's condition gradually improved. She regained some feeling in her right side and, with therapy, she was able to stand up and move her right leg. She couldn't walk, but she was definitely on the road to a partial recovery.

On October 9th, Mom was discharged from Casa Colina. I acquired the necessary equipment and made arrangements to take care of her at my home. I spent almost 24/7 with her. She was never alone. If I wasn't with her, a hired caregiver was.

During the next two months Mom's improvement was remarkable. By Christmas of 1998 she looked great. When she was dressed and made up, no one could tell she had suffered through a stroke. The only giveaway was that she couldn't say anyone's name, and she occasionally said the wrong words. She had a great sense of humor and those wrong words kept us constantly entertained.

I hadn't spent Christmas with Mom in 30 years and I was thrilled to have her with me. Although she couldn't walk, she continued to improve. I was so ecstatic I decided it was time for a holiday party. I knew most everyone on the guest list would have prior party commitments so I set the party date for December 27th.

That holiday season my home never looked better. I decorated it to the nines inside and out. I did it especially for Mom. I didn't know if it was going to be our last Christmas Season together and I wanted both of us to enjoy every minute of it. I hired caterers, a bartender, and booked the harpist who played for Mom's birthday party to entertain.

My brother Tom and his wife Carla drove from Illinois to spend the holidays with us. We were overjoyed and ready for a perfect holiday party.

The party was "Black-Tie," so that evening we dressed Mom in a gorgeous black and gold sequined and beaded dress. We bought her a corsage of red roses.

Mom always looked fifteen years younger than her age, but that night she was exceptionally beautiful. No one could tell she had been ill.

Over 75 guests attended the party. The guests included Virginia O'Brien, Jane Withers, and Margaret O'Brien. Margaret starred in "The Secret Garden," "Little Women," and won an Oscar for "Meet Me In St. Louis," co-staring Judy Garland.

I am a member of The Thalians Presidents Club. The Thalians is a Hollywood show-business charity chaired by Debbie Reynolds and Ruta Lee. Several of my Thalian friends, including Rudy Tronto, attended the party. Rudy has been in many Broadway shows. He also directed Ann Miller and Mickey Rooney in "Sugar Babies."

Mom always enjoyed talking to people in show business. That night she was in seventh heaven. Jane and Margaret sat near Mom and talked to her most of the night, as did my daughters, my brother, and his wife. Mom was the center of attention and she loved every minute of it.

I also invited Christina Miller to the party. Since our meeting in July, we had developed a good friendship and we were looking forward to the movie going into production. I was also anxious for Christina and Mom to meet, since Mom had also known Christina's deceased cousin, LaDonna.

Needless to say, the Christmas Party of 1998 was the most enjoyable holiday party I ever had. From the time the first guests arrived until the last guests walked out the door everything was beautiful, fun, and fulfilling.

CHAPTER 68

The New Year of 1999 started out to be very exciting. Kermit had been informed by the investors that we would have the money to start production on the film no later than January 20th.

Christina and I could hardly wait to get started. Each day seemed like a week. But I patiently spent my time with Mom and waited to hear the good news from Kermit.

I don't think I slept the night before January 20th. I was up early in the morning, eager to hear from Kermit. Every time the phone rang my heart skipped a beat. It was late that afternoon when I received Kermit's call.

"I hate to tell you this," Kermit said, "but everything fell through. The investors backed out on everything."

I felt sick inside. I couldn't believe the bottom had fallen out of everything again.

"How could that be?" I asked. "I thought it was a done deal, and they wouldn't back out."

"I don't know what happened," Kermit replied. "I'm just as surprised as you are. I thought everything was all set. I had no idea they would change their minds."

I knew Kermit was just as disappointed as I was, so I didn't ask any more questions. He felt bad enough. I didn't want to say anything that was going to make him feel worse.

"Well, it looks like we're batting zero," I said. "But don't feel bad. It isn't your fault. You did the best you could do."

"I'm not giving up," Kermit responded. "We may have to wait a while, but sooner or later we'll still get the movie made."

When I hung up the phone I was almost numb inside. I was upset and extremely disappointed, but I wasn't devastated. I had grown accustomed to false promises.

I couldn't be defeated by Kermit's news. I still had my script, and I could start over again. The thing I hated most was that I had to deliver the bad news to Mom and Christina. Christina was counting on being in my film, and Mom was elated that it was going to be made. I would rather have taken a beating than tell them everything had fallen through.

When I delivered the news they were disappointed, but neither of them was surprised. Christina had been involved with the entertainment industry. She knew not to count on anything until the papers are signed and the money is in your hand. Mom was deeply disappointed, but she was more concerned about my feelings than anything else.

I didn't know what to do. Not one person who made a promise to me had kept their word. I had coped with Bob Munger's false promises because I thought Kermit was getting money to produce the film. I thought everything was going to be ok. But that was not to be the case. My dreams had been shattered once more. I was at rock bottom and I had no one to help me.

For the first time, I seriously wondered what had transformed Bob from a supposedly devout "Born Again Christian" determined to make the movie, into an evil man who tried to destroy everything. Maybe he had never changed. Perhaps from the beginning his self-professed godliness was a facade. He had the chance to do what was right and he blew it. He failed his test. I knew some day, in some way, he would pay for it.

I was tempted to call Bob, but I knew my efforts would not bear good fruit. As far as I was concerned he was The Devil's Disciple and he had already done enough evil. I knew any further contact with him would only be inviting evil back into my life.

CHAPTER 69

In March of 1999, my sister Mary flew to California for a two-week visit with Mom and me. While she was here, another very disturbing thing happened. There was an article in the newspaper stating that Arnold Schwarzenegger was making a new movie called, "The End of Days."

That article stated he played the part of Jericho Cane. Jericho was the protector of a girl who held the future of mankind on her shoulders, and she was being pursued by the ultimate personification of evil.

I felt as though I was being electrocuted. That was the same plot that was in my script. I couldn't understand how someone else could have come up with my plot. Then I read the real kicker. Andrew Marlowe wrote the script. He was the screenwriter I had told my story to in the spring of 1998. Since I couldn't believe someone of his caliber would steal my idea, I had to find out what that movie was about.

By this time, Christina and I were like a brother and sister. I called her right away and told her about the article in the paper. Her reaction was the same as mine. She was just as upset as I was.

"Wait a minute," she said. "Let me call you back in a few minutes." About five minutes later my phone rang.

"Hi George," Christina said. "I used to work for an agency that handled scripts. I called there and just by chance they handled the script for 'The End of Days.' Some friends of mine still work there and we'll have a copy of the script tomorrow."

The next day, Christina and I picked up a copy of the script. When we looked at it, neither one of us could believe our eyes. The first thing we noticed was the name of the Production Company. "Lucifilms." That was so close to the name "Lucifer Films" that it was scary. I assumed "Lucifilms" was an acronym for "Lucifer Films." In any case, it's strange that a name similar to Lucifer was attached to the script.

When I opened the script and started to read it, I became more upset. The scenes were set up just the same as my script, and the characters were the same. But they had different names.

Then I saw the real kicker. The name of the lead girl was Christine Bethlehem. That was a Bob Munger name if I ever heard one. He had wanted me to change Amy's name to Christine and I had refused. There were many things in the script that resembled mine, but they were not all things I had not discussed with Andrew Marlowe.

When Andrew Marlowe told me he was working on a script with a plot similar to mine, I almost asked him if he was working with Bob Munger. Now I really wondered if Bob had a finger in the pie.

A few days after I read "The End of Days" script, I decided to call Bob Munger. When he answered the phone, I could immediately tell he wasn't overjoyed to hear my voice.

"Hello, Bob," I said. "This is George. How have you been?"

"Fine, and you," he replied.

"I'm fine too," I responded. "I'm just a little upset."

"What's the problem," he asked.

"Well, before I tell you what the problem is, I have to ask you a question. Did you ever show or give copies of my script to anyone?"

"No, why do you ask," he replied.

"Someone ripped off my idea and stole scenes from my script," I answered.

"How do you know that," he asked.

"Have you ever heard of the movie, 'The End of Days,' I asked.

"No," he answered.

"Well, there's a movie being shot right now called "The End of Days," starring Arnold Schwarzenegger, and it's a rip-off of my script."

"How did you find out about it?" he asked emphatically.

"There was an article in the paper. It said that Arnold Schwarzenegger was playing the role of a man who was the protector of a girl on whom the future of mankind rested, and she was being pursued by the ultimate personification of evil. That's the same plot my story has."

"That doesn't mean anything," Bob replied. "There's lots of movies made that had similar ideas."

"I know that," I responded. "But I read the script. There are so many things in it identical to my script that it's unbelievable. It starts out at the Vatican, like mine. It has the same characters, with different names, doing the same things and some of the scenes in it are almost identical to the scenes in my script. The most obvious thing is that "The End of Days" has the music box all the way through it, just the same as my story does."

"How did you get a copy of the script," he asked.

That's when I became suspicious. He showed no concern that my story or my idea may have been stolen. He only wanted to know how I found out about it.

"A friend of mine got me a copy of the script," I answered. "I had it the day after I saw the article in the paper."

"What paper was the article in," he asked.

"My local paper," I answered. "I'm filing a lawsuit. Someone stole my idea and used scenes that are identical to my script."

"Well, it isn't that easy to prove someone stole anything from you," Bob replied. "It's not easy to get an attorney to take a case like that. Most of them lose. Don't waste your money."

"I already have an attorney," I replied. "And he's taking the case on contingency."

When Bob heard that statement, his voice and attitude suddenly changed. I could tell he was surprised.

"What is your attorney's name," he asked.

That question made me even more suspicious. Why would he be concerned more about my attorney's name than he was that my story had been ripped off?

"I haven't decided which attorney is going to handle the case," I replied. "I've spoken with several, and they all said that it's obvious I've been ripped off. By the way, do you have a copy of the script Jim Hardiman wrote for me? I've misplaced my draft and I need another copy."

"I'm sure I have a copy around here someplace," Bob replied. "If I find it, I'll call you."

I intentionally lead Bob to believe I already had an attorney. I didn't think it was a good idea to let him know otherwise.

The following day I found Jim's draft of "The Devil's Reign." When I compared it to "The End of Days" I couldn't believe my eyes. Some scenes in the script written by Jim were not used in my final draft, but they were almost identical to scenes in "The End of Days." Bob Munger was the only person who had a copy of that script, other than Jim and I.

The film version of "The End of Days" was changed somewhat from the script, but similarities between their script and mine are as follows:

"The Devil's Reign," begins at the Vatican, then goes to years later, exactly like "The End of Days."

In "The Devil's Reign," Amy's father John is dead. In "The End of Days," Christine's father John is dead. (Bob Munger insisted that I change Amy's name to Christine. It's strange that Christine is the heroine in "The End of Days").

In "The Devil's Reign," Amy has a music box that was given to her by her dead father, John. In "The End of Days," Jericho has a music box that belonged to his dead daughter, Amy.

In "The Devil's Reign," a girl named Helen is involved in Amy's life when Amy is a young girl. Years later, Helen is Amy's boss. Helen is a disciple of the Devil, and Amy doesn't know it. In "The End of Days," a nurse named Mable is present when Christine is born. She removes Christine from the hospital nursery (a scene similar to a hospital nursery scene in Jim's draft) and has her life dedicated to the Devil. Years later, Mable is Christine's stepmother. Mable is a disciple of the Devil, and Christine doesn't know it.

In "The Devil's Reign," (Jim's draft) figures emerge from the sewers. In "The End of Days," figures emerge from the sewers.

In "The Devil's Reign," Amy is hanging from a balcony and the music box is playing in the background. In "The End of Days," (script) Jericho is hanging from a fire escape railing and the music box is playing in the background.

In "The Devil's Reign," Amy goes to the cemetery and places flowers on a grave. She looks up and sees an old homeless looking man. Later, we discover that man is evil. In "The End of Days," (script) Christine goes to the cemetery and places flowers on a grave. She looks up and sees an old woman. Later, we discover that woman is evil.

In "The Devil's Reign," God gives Amy her destiny. In "The End of Days," the Devil gives Christine her destiny.

In "The Devil's Reign," Helen tries to kill Amy, and Helen is killed. In "The End of Days," Mable tries to kill Christine, and Mable is killed.

At the end of "The Devil's Reign," the Devil (through Helen) has control of Amy's friend and tries to get him to kill Amy. At the end of "The End of Days," the Devil had control of Jericho's friend and tries to get him to kill Jericho. In both scripts, the Devil loses.

At the end of both scripts there are figures of saints with swords, and the swords have an important impact on the story.

My story is based on fact. It is a godly story containing a message that, hopefully, will change the lives of some people. "The End of Days" is a fictitious demonic story that serves no purpose.

Bob Munger never once called to express concern over what had happened to my story, and he never sent me a copy of Jim's script.

Andrew Marlowe has to know the plot for "The End of Days," is not original. Every word of my script and my story is original and every statement I have made in this book is true. I have a paper trail that goes all the way back to 1983. I can validate everything I've written with legal documentation, copyrights, and witnesses. I will also gladly submit to a lie detector test. Are Bob Munger and Andrew Marlowe willing to do that?

Bob had no reason to be angry because I struggled to save my story, or because Tom and I had rewritten the script. I believe he ripped my script apart and was angered when I told him I wouldn't re-write it so he had an excuse to walk out. I believe he already had his own deal going with "The End of Days."

I believe Bob wanted to steal my story from the beginning. That's why he tried to get Tom to change my story to his liking, why he purposely deceived me, and why he never gave Jim a copy of my script or treatment.

An interesting thing I discovered is that Bob's first name, Robert, has 6 letters in it. When you break it down to numbers it equals 6. His last name, Munger, also has 6 letters in it. When you break it down to numbers it equals 6—"6666." A coincidence? Maybe so. Maybe not.

When I did the numerology on the 1972 Ford with the "LUCIFUR" license plate, everything added up to 1998 and 666. At that time I said a major event involving the Devil was going to take place in 1998. I didn't realize that in 1998, Bob Munger would do the Devil's dirty work and destroy my story. Was it a coincidence? Possibly. But I don't think so.

I talked to several attorneys. But the ones who wanted to help me weren't experienced in the entertainment field. The others wanted too much money to handle the case. I was told it would cost at least two million dollars to litigate the case because of the time it would take for investigation. I was also told I had to prove the connection between Bob and "The End of Days." I think there were other reasons the attorneys didn't want to handle my case. I felt they didn't want to find a reason to file a suit. At least I was told that the statute of limitations didn't start until I had proof Bob was the culprit.

I tried to get the media interested in my story, but my efforts were fruitless. No one cared about my story, my mission, or me. No one would help.

Several years earlier, a writer for the local newspaper wanted to write a large article about my ordeal. She wanted to tell the entire story, and include a photograph of the search warrant and me. She was excited about writing the article. However, when she presented the idea to her editor, she was told, "You can't do it. We don't do stories about God. They belong in a religious publication."

The writer was extremely disappointed and apologetic that she had to cancel the article. I was upset, but I was even more disturbed about a week later. The same newspaper ran a large article about Devil worshipers. Obviously the media considers the works of the Devil and the deeds of evil people to be more newsworthy. God and the deeds of godly people are seldom mentioned.

CHAPTER 70

Greedy and phony people in Hollywood had shattered all my hopes and dreams. Only a few weeks before, I was walking on clouds. Now I was at rock bottom and further in debt. And I had to watch other people make a fortune from my idea.

I guess the Devil thought I wasn't hurting enough. In the January 19-25, 1999 issue, the Hollywood Reporter published an article announcing the production start date for "The End of Days." The article stated that the Production Company was "Lucifilms" (Lucifer?) and the Production Start Date was, of all days, November 21, 1998—my birthday.

Before this book went to press, I called the legal department of the Hollywood Reporter and requested permission to print a copy of that announcement in this book. It is beyond me why my request was denied. Instead of presenting a copy of that Hollywood Reporter "Films in Production" page for documentation purposes, the original is on file.

The Devil had kicked me in the teeth and stomped me into the ground. No one cared about the truth and no one would help me. But I couldn't give up. I couldn't let the Devil win. I was back on home plate and I was batting zero, but I wasn't out. I still possessed two things the Devil could never destroy: my faith in God and the determination to fulfill my mission.

Bob Munger took my twenty thousand dollars, wrecked my script and stole my idea. But he couldn't keep me from writing a book and telling the true story. And he had given me a new chapter to add to it.

Within days I started working on this book. For a change, words seemed to flow easily, and in less than two weeks I had written over forty pages. I wanted to have the book completed before "The End of Days" was released.

Business was good at the stores and I was making progress on my book, then I received more bad news. A large bridal outlet was going to open only a few miles from my bridal-tuxedo shop. My bridal business was booming, but competition was heavy and I knew I couldn't compete with an outlet store backed by lots of money.

My lease was going to expire in July and my landlord was going to raise my rent, so I thought it was best for me to get out of the bridal business before it was too late.

My current lease didn't expire until the end of June, so I had a few months to close the bridal business and move the tuxedo store to a smaller location. I was going to have plenty of time to make the transition. That is until I heard even more bad news. A tuxedo store, owned by a major chain associated with the bridal outlet, was going to be opening in the area. A second competitor was already in the process of opening another nearby tuxedo store.. I had to make changes fast. I decided to offset the threat of competition by opening a new store in Riverside. Needless to say, I had little free time, and I had to stop working on the book.

I signed the lease for the Riverside location the first week in February. I hired Danny DaValle, a friend from Chino, California, to construct the interior of the store. Danny had previously taken possession of my apartments and had made a promise to buy the building on contract for deed.

The store was supposed to open in two weeks. But Danny didn't do as he had promised, and the store didn't open for three months. I paid three months expenses and lost three months' income, all during what would have been the busiest months of the year.

Mom's health continued to improve, and her attitude was wonderful. I hired a caregiver to stay with her during the day, and I stayed with her every night. I planned to build her a room in the back of the store. Then, if she felt good I could take her to work with me. Unfortunately, the store was open for only a few weeks before Mom's health started to fail.

I returned home from work one evening to find Mom sweating and short of breath. I thought perhaps the house was too warm, but when I turned on the air conditioner, it made no difference. Late that night I took her to the emergency department at Kaiser's Fontana Hospital. According to the doctor, her lungs were clear and she only had minor congestion. Mom didn't look good to me, but he said she would be fine. So I took her home and put her to bed.

The following evening I returned from work to find Mom's condition hadn't improved. The congestion in her lungs seemed to be worse, but I wasn't concerned since the doctor said she was going to be fine. Mom had several severe coughing spells, but she managed to sit up and talk until bedtime.

When she was ready to retire I helped her into her wheelchair, pushed her into her bedroom, and got her ready for bed. The second she laid down, I knew something was wrong. She started gasping for air and she panicked.

"Get me up, get me up quick," she yelled frantically. "I can't breathe."

I helped Mom sit up. She was more comfortable, but she still complained that something was wrong.

"Mom, I don't know what to do." I said. "I just took you to emergency last night. The doctor said you only have a little congestion and you should be fine. You need to give yourself time to recuperate."

"We need to go back to the doctor," Mom insisted.

I thought going back to the emergency room was pointless, but I knew I would be up most of the night if I didn't appease her.

I got Mom out of bed and we went back to the emergency room. A different doctor examined her. He was concerned about her condition, and he admitted her to the hospital. I felt like the scum of the earth because I had been impatient with Mom when she wanted me to bring her back to the hospital. While the papers for admission were being filled out, I stood beside her and held her hand.

"Mom, please forgive me for being short with you," I said. "I had no idea you needed to be in the hospital. I love you with every bone in my body and would never purposely do anything to hurt you."

Mom lay on the gurney and gasped for air. "It's ok, I know that," she replied. "I'm thirsty. Can you get me a drink?"

I gave Mom a drink of water and less than a minute later, she started coughing and the water came pouring out of her mouth. I thought it had gone down her windpipe. I told the doctor what had happened, but he wasn't concerned. He thought she had regurgitated the water. Soon after that, she was taken to her room and started receiving a respiratory treatment.

"I think your mother is going to be fine," the doctor said. "This treatment will help clear her lungs, and the medication should help her get some rest."

Mom seemed to be contented and relaxed, so I said goodnight to her and went home to sleep.

The following evening, I went straight from work to the hospital and rushed to Mom's room. I expected to see her feeling much better and anxious to go home. Instead, I was confronted with devastating news. They were unable to clear her lungs and she had been placed on a respirator. I was heartsick and felt even worse about being impatient with her the previous night.

Mom was lying in bed with a ventilator tube down her throat. She was miserable. I stood beside her bed, held her hand, and prayed. Even with her misery, she managed to look into my eyes and nod. It was all I could to do to keep from crying.

"Mom," I said quietly. "I had no idea you were so ill. Please forgive me for being impatient with you last night."

Mom squeezed my hand. I knew she was telling me everything was fine and not to worry about it, but I would have given anything to turn back the clock and have been more compassionate.

The next day my sisters flew to California. From the moment they arrived, we were at Mom's bedside. She couldn't speak, but with a nod she acknowledged everything we said to her. Her kids were her life, and she was ours. Even through the mask of misery that covered her face, we could see she was happy to have three of us with her.

Mom's condition began to improve and her lungs cleared, but the doctors were having a difficult time weaning her from the ventilator. They removed the tube from her throat, performed a tracheotomy, and placed the tube in her neck. She was a lot more comfortable, but she still couldn't talk.

After several more unsuccessful attempts to remove her from the ventilator, Mom was moved to a rehabilitation hospital in Pomona. Her condition continued to improve, and we prayed she would soon be weaned from the ventilator and go home. Every attempt was unsuccessful, but her therapists remained optimistic that they would eventually succeed.

Mary and Jean stayed for three more weeks before they returned to Illinois. I went back to work, but every evening I went straight to the hospital. Mom was my priority.

Many times I walked past the nurses' station on my way to Mom's room. Quite often someone at the desk would say to me, "I can't believe you're here to see your mom every day. Most people don't come to see their parents once a month."

I was saddened to hear that. I wanted to see my mother every day. Many days I would visit her twice. When I was unable to visit her, I felt guilty. I couldn't understand why other people didn't care just as much for their parents.

Mom had always taken pride in how she looked, but she no longer wore her false teeth and the nurses kept her hair combed into a little ponytail on top of her head. She wasn't the classy lady she would have liked to be. But she was an adorable happy little round-faced granny. When she smiled, she was as cute as a button.

Mom was the hospital staff's favorite patient. They nicknamed her "Smileymom," because of the smile she gave everyone who entered her room. I was rewarded every time I went to see her. Mom couldn't talk, but the moment she saw me her eyes lit up and a big smile covered her face. She always motioned for me to hold her hand.

Every time I stood beside her bed my heart ached. I constantly prayed that she would soon recover and I could take her home with me.

CHAPTER 71

I didn't have time to work on my book, and I didn't have time to close the bridal shop and move the Chino tuxedo store. But, I found a smaller and better location across the street, and Randy took care of everything else. He worked relentlessly. It's a good thing he wasn't getting paid by the hour, or the stores would have gone broke.

Before I realized it, it was October. The bridal shop was a thing of the past and the new Chino Tuxedo Junction store had been open for over a month. Business was ok, but Mom's health had started to decline.

During the past few months Mary and Jean had been out to see Mom several times. My brothers were just as concerned about Mom as the rest of us, but they had to work and it wasn't as easy for them to get away. In mid- October, they could no longer wait to see her, and they flew to California.

I met Tom and Ed at the airport and took them directly to the hospital. When we walked into Mom's room, as usual, her eyes lit up and a big smile covered her face. But this time was different. She was thrilled to see them and she never took her eyes off them for a second. She looked back and forth from one to the other and smiled and winked at them, then she puckered her lips as if she was giving them a kiss. She couldn't utter a word, but we knew every thing she was trying to say.

Tom and Ed talked to Mom for a while, and then they made an excuse to leave the room for a few minutes. It had torn them apart to see Mom in such pitiful condition. We all three stood in the hallway outside her door and cried. It killed us to see the mother who had raised us so unselfishly and loved us so unconditionally, hooked to a machine, suffering, and unable to talk. Yet she smiled, and without saying a word she let us know she loved us very much.

Tom and Ed were only able to stay for two days, but it was two days that Mom relished. Every moment they were at her bedside, her eyes never left them. Several times, tears welled up in her eyes and rolled down her cheeks, and she silently cried.

The week after Tom and Ed flew home, Mom's condition suddenly improved. Every time I walked into her room she looked at me with a big smile and motioned for me to hold her hand. To my surprise, even with the tracheotomy, she managed to say, "Thirsty."

I was so thrilled to see her doing so much better I got her a Dr. Pepper. She slowly sipped every drop of it through a straw, and she loved it. For the next few days, as soon as I walked into her room, she was waiting for me to get her another Dr. Pepper. On one occasion, as soon as her thirst was satisfied, she even managed to say, "Good."

I couldn't believe how much she had improved. Except for the ventilator, she looked like she was in perfect health. My hope was renewed that she would still get off the ventilator and eventually totally recover. I planned to turn my dining room into a hospital room, and I hoped to have her home by Christmas.

On the fourth day of Mom's rally, Mary flew back to California to see her. Mary was amazed that Mom looked and felt so much better. She was alert, energetic, and the change in her overall appearance was remarkable.

The fifth day, Mary and I went to see Mom and she still looked wonderful. Then, just as fast as she had rallied, she had a relapse. Our hopes for Mom's recovery were shattered.

Mary and I stood by Mom's bedside and watched her suffer. She was perspiring heavily and gasping for air. Her face was red and her eyes were filled with pain. Yet, she tried to tell us something. She grabbed Mary's hand and squeezed it tight, and she looked back and forth at the two of us.

I placed my hand on Mom's knee and looked into her eyes. "I love you, Mom," I said softly.

She gave me a nod that acknowledged the feeling was mutual.

Mary was frightened, but I had seen Mom suffer through dozens of similar episodes. The therapist always adjusted the oxygen level on the ventilator and soon she would recover. I knew, within an hour, Mom would be fine.

A nurse's aide came into the room and asked Mary and me to wait out in the hallway so she could change Mom's bed. While we waited, the respiratory therapist spoke to us about Mom's condition.

"There's nothing to be concerned about," he said. "She's going to be fine."

"I don't know," Mary replied. "Her condition's deteriorated, and I'm really concerned about her."

"I've seen Mom bounce back from episodes worse than this one," I said, hoping to comfort Mary. "You haven't been here to see it happen, so you don't know what to expect."

"That's right," the therapist replied. "He knows what I'm talking about. He's seen her go through this many times. Believe me, she'll be fine."

It was after eight o'clock, and it was going to be a while before we could go back into Mom's room, so the therapist convinced Mary and me to go home and relax.

"Will you call us if Mom doesn't improve," Mary asked. "Then we'll come back."

"Yes, of course I will," the therapist responded.

"Even if she's better, call us at ten-thirty," I added. "We want to know how she is doing, either way."

"No problem," he replied. "I'll call you at ten-thirty."

At ten-thirty, as promised, we received a call from the therapist. Mom was doing fine. Her breathing was normal and she was sleeping. Mary and I breathed a sigh of relief and went to bed.

CHAPTER 72

At six-thirty the next morning the phone awakened me. I was afraid of what I might hear when I answered it.

"Is this George Newberry," the man asked.

"Yes it is," I replied.

"I hate to tell you this, but your mother is deceased."

"Oh my God," I responded, "we'll be right there."

I didn't give the man time to say more. I slammed the receiver on the hook and headed for Mary's room. I was so shaken I could barely climb the stairs.

"Mary," I screamed, "Mom is dead, our Mom is dead!"

Mary instantly shot out of her room, already hysterically crying. We were both in a state of shock, but within minutes we were dressed and on our way to the hospital.

We arrived at the hospital and hurried down the corridor towards Mom's room.

"I don't know if I'm going to be able to handle this," I said to Mary.

"We have to be strong," she replied. "Mom made us promise that when she passed away, we wouldn't cry for her. She made her peace with God a long time ago, and she was ready to go."

"I know," I responded, "but I can't imagine living without her. My life will never be the same."

We entered Mom's room, and the second I looked at her lying lifeless on the bed, I fell to pieces. I walked to her bedside, held onto the bed rail, and wept. Then I noticed a smile on Mom's face.

"Look, Mary," I said, with a quiver in my voice. "Mom even died with a smile on her face."

"Maybe that's her way of letting us know she's with God and she's happy," Mary replied, struggling to control her emotions. "Now she won't have to suffer any more."

"I know she's with God," I replied. "She was a wonderful mother and that's the only place she can be. But I'll miss her 'til the day I die."

I leaned over the bed rail, and kissed Mom on the cheek. Then I held onto the bed rail and cried. Mary stood on the opposite side of the bed and did the same.

"We love you, Mom," Mary whispered, as she lovingly stroked her fingers through Mom's hair and rubbed her cheek. "You're in heaven now. Say hello to Gramps for me. Before long we'll all be together again."

We only stayed with Mom a few more minutes. There was no more we could do for her. We were only punishing ourselves by remaining at her bedside.

That afternoon, Jean flew out from Illinois and the three of us started making funeral arrangements. We decided to have two services; one in California and one in Illinois.

The evening of the California service, I had a hard time walking up to Mom's coffin. But when I saw her face, I felt much better. She couldn't have looked more beautiful. She looked like a movie star, and she still had that smile on her face. She was at peace.

The first funeral service was held in Ontario, so California friends and relatives could attend. Then we flew with Mom to Illinois for the second service.

When the plane taxied down the runway and the wheels lifted off the ground, I cried. My mother was leaving California for the last time. I knew I would never have her back with me again.

Before the service in Illinois began, Mom's open coffin was sitting inside the entrance to the church. My siblings and I stood beside her and we greeted people as they entered. I remained beside Mom's coffin until everyone was seated inside the church. Then, I leaned over and gave her one last kiss on her cheek.

Mom had many friends and relatives in both California and Illinois, and a large crowd attended both services. Also at both services, just as at my stepfather's funeral, the tape of my song "Those Were the Times to Remember" was played, and one of Mom's favorite songs, "How Great Thou Art," was sung. I listened to the words of both songs and I cried.

I had always prayed that Mom would live to see my mission fulfilled. But that was not meant to happen. God had other plans for her and me.

Half the joy of my having a successful book or movie was gone. I would never be able to share my success with my mother. But I couldn't stop. I had to go forward. and fulfill my mission. I knew if I ever succeeded, Mom would still know it.

CHAPTER 73

I dealt with a lot of disappointments and pain in 1999. The funding for my film fell through, the plot of my script was stolen, my mother passed away, and "The End of Days" was released.

"Lucifilms" produced "The End of Days," and the production start date was November 21, 1998—my birthday. I thought it was ironic that the film was released during the week of my birthday in November of 1999.

I was dealing with a lot of disappointments and heartaches, but I knew I would gain nothing by wallowing in my misery. Even though my prayers hadn't been answered, I knew God had His reasons. Perhaps it was to make me a wiser and stronger person. Then too, with every new experience, I had a new chapter to add to my book.

Every experience strengthened my faith in God and made me more determined to fulfill my mission. I had a story to tell and I was resolved to finish writing it. My Uncle Earl set up a computer in the back of the Riverside store, and every minute I had to spare, I worked on my book.

Months passed. Words were flowing easily and I made excellent progress on the book. But my problems weren't over. Danny DaValle, who had taken possession of my apartments, had been collecting the rents but he had not been making the mortgage payments.

I called Danny. He told me it was all a mistake and he was getting it all straightened out. He even talked to the mortgage companies, on conference calls with me, and told them the same story. Freddie Mac was already involved. Danny swore he had made the payments and he was going to send them copies of his canceled checks.

It wasn't long until I received another notice from the mortgage company. They still hadn't received any confirmation that the payments had been made.

I called Danny again.

"Hey Danny, what's going on?" I asked. "I got another notice from the mortgage company. You still haven't sent them any confirmation that you've made any payments on the apartments."

"I don't understand what's going on," Danny replied. "I sent them everything they needed. Everything is going to be fine. Don't worry about it. I'll take care of it."

Within days I got a notice that the apartments were going to be sold at a foreclosure auction.

That evening I went to dinner with Bob and Cindy Mills. They knew Danny very well and, like me, they were suspicious that Danny wasn't telling the truth. I called Danny from my cell phone.

"Danny, what are you doing to me?" I asked. "Tell me if you aren't telling me the truth. I've got to get the apartments out of foreclosure."

"George," Danny replied. "I'm telling you the truth. I wouldn't lie to you. You are my friend."

"Then what's going on with the mortgage payments?"

"I found out the envelope I sent them in got burned up in that mail truck that caught on fire by Palmdale, California." He replied. "Everything will be straightened out. I promise."

Bob and Cindy were listening to most of what Danny said to me. A mail truck had burned near Palmdale, but there was no way a letter from Chino was going to be on it. We knew something was screwy.

A few days later, Danny called me. He told me he had taken care of everything. There would be no more problems. He faxed me a copy of a check for $5,000 he had sent to a foreclosure company. He also sent a copy of a letter from the foreclosure company stating that upon receipt of the check, foreclosure proceeding would be stopped.

Danny's explanations to the mortgage companies were lies. And, the copies of the check and letter he faxed to me were phony. I didn't find out he wasn't telling the truth until a month after my building was sold in foreclosure.

I will never understand why he didn't want me to know the truth. If he had been honest with me, I could have saved the building. But as a result, he destroyed my credit rating, I lost the building, and the matter had to be settled in court.

I was awarded a judgment against Danny. But he hid all his assets in his wife's family trust, and claimed he didn't work. I was left with nothing but a piece of paper. He drives a new truck, lives in a million-dollar house, and is a property manager for his wife's family.

In the past, I would have been emotionally destroyed by the result of Danny's actions. But, I have been confronted by so many dishonest people and have dealt with so much despair I have become a much stronger person. Besides, I know where the problems are coming from. As long as I'm true to my mission, I know the problems will not subside. It's an ongoing battle anyone who is fighting evil must confront. I have to keep my faith in God, and I can't let the Devil destroy me. I will win.

CHAPTER 74

While I was working on my script, I called an old friend of mine, Rita Reber, who now lives in Florida. Rita and I grew up together in Illinois and I am Godfather to her daughter, Cherese, who lives in Longview, Texas. I told Rita about my goal and the story I wanted to tell. In return, Rita told me something that proved to be very interesting.

"Have you heard about the apparitions that are occurring in Medjudgorie, Yugoslavia," Rita asked.

"No, I haven't," I replied. "What kind of apparitions?"

"They're similar to the ones that took place during the Miracle of Fatima," Rita responded.

For those of you who don't know about the Miracle of Fatima, it occurred in Portugal in 1917. The Virgin Mary appeared to three children, giving them messages to relay to the world. The children were scorned and scoffed at by many people and they were even imprisoned by their government. But in the end, on October the 13th, 1917, 70,000 rain-soaked people joined the children in a "cova," and they witnessed the miracle that proved the children were telling the truth. The rain stopped, the clouds parted, and the sun began to spin in the sky. The people were in awe. Suddenly the sun started to plunge to earth and the horrified people began to run for shelter, some thinking it was the end of the world. As suddenly as the sun started to fall from the sky, it rose back to its original position, and radiated with beams of color. The sky cleared, and—miraculously—people and the earth were dry. The lame could walk, and the blind could see.

Newspapers of that time contain records of that event, and many people who witnessed the miracle are still living.

The prophecies of Fatima were fulfilled. Now there are countless books that deal with the subject, and in 1952 a movie was made, entitled, "The Miracle of Our Lady of Fatima." I saw that movie when I was twelve years old. It touched my heart and changed my life. I became a very spiritual person. Because of that experience, I know that films do have an influence on people's lives.

"Rita suggested that I order some of the books written about Medjudgorie. "The messages being given during those apparitions coincide with what you're saying," she said. "I think you will find them to be very interesting."

I ordered books and videotapes that told all about the apparitions in Medjudgorie. The first apparition occurred in 1981 and they have continued up to the present. When I was reading one of the books, I found that one message given to the visionaries was especially interesting. It basically said:

God and the Devil conversed and the Devil told God, "People only believe in you when things are good. When things are bad, some people cease to believe in you and others act as if you don't exist. Therefore, God gave the Devil one hundred years to have more power over the world and the Devil chose the Twentieth Century."

I don't know that the apparitions in Medjudgorie are authentic, but I found that message to be quite enlightening. Especially since it coincided with the basic message of my story.

Another message that caught my attention said, "God did not separate men by religions. Men did it to themselves. A person cannot be a good Christian if they make fun of other religions; for people of all religions will go to heaven."

That message was another confirmation of my beliefs. As I said before, I do not believe God would keep good spiritual people out of heaven just because they weren't Christians.

Recently Pope John Paul said. "All who live a just life will be saved, even if they aren't Christians and they do not believe in the Roman Catholic Church. The Gospel teaches us that those who live in accordance with the Beatitudes – the poor in spirit, the pure of heart, those who bear lovingly the sufferings of life – will enter God's Kingdom."

I was pleased that Pope John Paul's statements supported my own beliefs. I do believe Christ was the Son of God and that He died on the cross for the forgiveness of our sins. Christians have Christ, God and their deeds; non-Christians have God and/or their deeds. Christ died for the forgiveness of everyone's sins.

All of the messages given at Medjudgorie were in accordance with the message I wanted to relay to people via my story.

The messages repeatedly said, "Pray, pray, pray, and get your lives in order with God, for time is short."

Another profound message said, "The Devil loves to destroy strong believers, and he easily influences weak people to live evil lives."

That message really hit home. I know the Devil would love to destroy me. I am his enemy. Even without apparitions or miracles, it should be easy for people to see that weak people are easy prey for the Devil. Young people are especially vulnerable.

Many young people are weak and are easily influenced by people who entice them to do evil. But young people are not alone. There are many weak adults too. When life isn't easy, or because of greed for money, power and drugs, it becomes easier for the Devil to influence people to live evil lives.

CHAPTER 75

As I said earlier, "Drugs are the Devil's keys to people's souls." Every day, more innocent people are sucked into the world of drugs by gangbangers, drug dealers, or friends who are already addicted.

At first, the drugs are free and the tempters try to convince their victims that it's harmless to try them. Weak people who succumb to the pressure of their tempter gradually become dependent on the drugs and, before they realize what's happened to them, they are addicted.

As people become addicted to drugs, they gradually lose their conscience. Once their conscience is gone, they like the way they are and they are easily influenced to do most anything. They will lie, deceive, cheat and steal from anyone, including their families and people they love. Some can even commit murder. With no conscience, they have no desire to change their lives, and the Devil owns their souls. They become the tempters, or Devil's disciples, and they join the ranks of those who help win souls for Lucifer.

The Devil's disciples help do his dirty work and try to spread evil everywhere. They walk among us. They confront us almost every day of our lives. But it is impossible to recognize them by their looks. They look no different than you or me. They may be our friend, neighbor, relative, subordinate, advisor, leader, or even our minister. They may be someone who we've never met, like a celebrity or rapper using their influence on their fans to promote evil.

They are sly. Many of them hide their evil intentions behind facades of kindness, love and friendship, in order to win the trust and respect of their victims. They must attain power over them.

False Prophets possess one trait they can't hide. That trait makes it easy for us to recognize them for what they really are. A disciple of the Devil will always try to convince you to think or do something that is ungodly and evil. The moment they try, you know they're doing the work of the Devil. They may truly be your friends, and they may truly care for you. They may not even realize they're working for the Devil. The Devil uses many people without their knowledge. But that's when you must take control of your own life and make them realize what they're doing. Then, you not only save yourself from the clutches of evil, you also have a positive influence on their lives, and you may help them save their souls.

Drugs have no prejudice. They will destroy anyone. Age, race, or social status makes no difference. They have obliterated the lives of weak young people and educated adults, as well as people entrusted to uphold and enforce the law, and government leaders. They have destroyed neighborhoods, cities, and even entire countries. Yet, government leaders sit passively by and watch as evil people work double duty for the Devil. Until the punishment for a crime is just as bad or worse than the crime, crime pays. If stricter laws mandating harsher punishments aren't passed and enforced; drugs, gangs, corruption and evil will, in time, destroy the world.

People should be able to see most of the signs indicating that an evil force is trying to control the world. It doesn't take a genius to spot them. I have mentioned the invasion of drugs and how they have affected the world. But there are numerous other venues the Devil is using to destroy mankind.

Some of them are obvious and some are not so easy to see. I'm sure many of them are hidden from the general population of the world and are only known by their perpetrators. If we were aware of every intrusion of evil that has or will affect mankind, we would probably not be able to sleep. I will briefly mention some of the signs that may have been inconspicuous to some people.

CHAPTER 76

Some atheists, who I believe to be disciples of the Devil, have waged war on God. If they win, God will be thrown out of the United States. Sadly, the courts have already granted some of their demands. They have succeeded in getting prayer removed from schools. If they have their way, prayers will never be spoken in any government-owned building or on any government-owned property. And, the word "God" will be removed from all government buildings, the Pledge of Allegiance, and government currency. Unfortunately, because atheists are offended by any reference to God or organized religions, our government is gradually kowtowing to their demands.

I live in Ontario, California. For the past 40 years, during the Christmas season, 12 hand-carved nativity scenes have stood on the median of the main downtown street, Euclid Avenue. Because of complaints filed by a local atheist activist, the future of the nativity scenes was in jeopardy. The activist demanded that no city funds could be used to store or move the twelve scenes. His demands were granted. But the citizens of Ontario united and found a solution to the problem. Money for maintenance and storage space for the twelve scenes was donated.

The activist still wasn't appeased. He didn't want the nativity scenes sitting on the main street median at Christmas time.

Once more people were outraged by his relentless demands, and they fought back. The atheist finally realized people were not going to give up the nativity scenes without a fight. He finally relented and agreed to drop his complaint.

But like the Devil he works for, his word was no good. He filed a complaint with Caltrans, a state highway agency. The City of Ontario was notified that only four of the nativity scenes could be placed on the median. And, they were to be joined by displays from other religions. The reason Caltrans supposedly had the right to dictate their demands to the City of Ontario was because Euclid Avenue is legally a state highway.

Once again the people fought back with the help of the Chamber of Commerce, a State Representative, a State Senator, and attorneys. Their position was that the nativity scenes were expressing freedom of speech. Caltrans relented and agreed to allow all of the nativity scenes to be displayed. But the atheist activist vowed to continue his fight. Needless to say, he is probably the most despised man in the valley.

I agree that other religions should be able to have displays on the median. But if people like the atheist activist have their way, eventually religious books will have to be removed from public libraries and religious works of art will be taken from public museums. Yet symbols of evil can be posted anywhere and those same people never complain. If this country bends to the demands of ungodly people, and God is removed from everything this country stands for, we are doomed.

CHAPTER 77

One of the most powerful organizations in America is the American Civil Liberties Union or, as it is commonly called, the ACLU. To me, the ACLU is the right arm of the Devil.

The ACLU, backing the demands of atheists, fought to get prayer out of schools. They fight for the right to be ungodly and they fight to do away with any rules based on spiritual beliefs. The ACLU doesn't fight for the rights of people who want to live godly lives, and they don't fight for the rights of victims of evil acts. They fight for the rights of people who choose to be ungodly and for the rights of criminals and perpetrators of evil. They are ready to file a lawsuit against any person or establishment that strikes back at someone who is committing a crime or evil act.

Whether the criminal robbed, shot, or murdered someone isn't the issue that concerns the ACLU or their followers. All they care about is whether the law, the victim, or law enforcement treated the criminals with respect and didn't infringe on their civil rights.

I believe, when someone is committing a violent crime he or she should have no civil rights. They should be held responsible for anything that happens to them. The laws and the ACLU should protect victims—not criminals.

How does the ACLU, or anyone, expect people who are being victimized to remain calm and think rationally? Even if people are trained law enforcement professionals, they shouldn't be expected to dodge a barrage of bullets and have sympathy for the aggressors. It is the normal human instinct for one to be angered or frightened by a violent aggressor. When a person chooses to commit a crime, he or she should accept the consequences of their actions. It is not the responsibility of their victims. People who want their civil rights protected shouldn't infringe on the rights of others.

One day when I was thinking about the ACLU, I decided to do numerology on its name—the same as I did with the 1972 Ford and Lucifur license plate that appeared on the search warrant. On the search warrant, when I added 1972 with the numbers for Ford, 6694, the result was "8666". I was astounded to find that the numbers for the ACLU added up to "6668", the reverse of the 1972

Ford. That could be a coincidence. But I don't think so. (A copy of the numerological chart I did on the ACLU is included with other documentation at the back of this book.)

Some of the actions of the ACLU have been commendable. But, they must do some good to dupe people into thinking they are here for our benefit.

The ACLU has succeeded in getting prayer out of schools. They protect criminals and fight laws that can help deter crime. Parents and teachers no longer have the right to discipline unruly, belligerent children. Evil children protected by unjust laws now control their parents, and gangs flourish because the law does nothing to curtail their existence. The world is in moral decline.

For centuries the rod wasn't spared. When I was a child, my mother used it, and on more than one occasion I had marks on my butt. My mother was not a cruel woman, and those marks didn't hurt me.

I grew up, like most children of my generation, having respect for my parents and other people. We called adults "Mr., Mrs., or Miss." We always asked, "May I" and answered, "Thank you, Ma'am," or "Thank you, Sir." We didn't talk with food in our mouths, and we didn't open our mouths or smack our lips when we were chewing food. We weren't rich, but we were taught to have manners, class, and respect for our superiors.

Now many young people are never introduced to culture or class, and many of them grow up with little or no discipline. To make matters worse, some of them are attracted to friends or celebrities who are negative role models. As a result, many young people are reflections of the negative people who influenced their lives. They have no manners and no class, and they respect no one but themselves and their peers.

The ACLU and the "experts" have destroyed everything our ancestors learned from thousands of years of experience. We placed our trust in educated professionals and elected political leaders. They call themselves "experts." We trusted the "experts" to resolve our problems. Instead, those "experts" have created a living nightmare. We don't have to be experts to see that the "experts" have the world in a mess.

Now, parents who use the rod can be jailed for child abuse. I believe that is another one of the Devil's schemes to corrupt the world.

He's taking control away from responsible adults and giving it to weak young people who are easily influenced to do evil. Laws protecting juvenile criminals should be done away with. Juveniles, especially teenagers, know when their actions are right or wrong, just as people of other generations did when they were children. Age is no excuse.

Until the last few decades, violent juvenile crime was a rarity. People feared adults more than they feared kids. Now, it's the other way around. Some of the most violent criminals are young people. They have no respect or compassion for other people and they most likely will live a lifetime of crime.

I don't mean to insinuate that all children are weak and succumb to every temptation to commit an evil act. That is not so. There are more good children than there are bad. But organized gangs of weak young people have succumbed to the will of the Devil and are aiding his quest to destroy the world.

As I have said before, "Drugs are the Devil's keys to people's souls." Most of the crimes committed by young people are drug-related, and most young drug-addicted criminals have no conscience. They rob, torture, and even murder other people, yet they feel proud of their actions. That's because they're no longer capable of feeling remorse and in the eyes of their peers, they are heroes.

Too many people escape penalties for their actions because unjust laws protect them. The penalty for each specific crime should be the same for everyone. The penalties shouldn't be based on mitigating circumstances, and they shouldn't change from one court, or one state, to another. For example: A robbery is a robbery and a murder is a murder under any circumstance. But we basically live in fifty different countries. A person can be convicted of a crime in one state for something that isn't even illegal in another. And in some courts, people have received a harsher sentence for drug dealing, robbery or lesser crimes, than others have received for committing murders.

It doesn't take experts to see what's wrong with the world. The experts got us in this mess. Many people see things that are wrong with their own governments, just as their government leaders see the things that are wrong with other governments. Yet the government leaders try to straighten out other governments and do nothing to make things right in their own countries.

Corrupt government leaders scam governments, get kickbacks, overpay themselves, waste money on projects that are absurd, and on and on and on. Yet, seldom is anything done to stop the corruption or the waste of money that belongs to the people.

Many politicians only take care of themselves and their governments. They only care about power, money and how they can take more from the people. They don't care about the working citizens who pay all the bills. Even in the United States, what was once a government of the people, by the people, and for the people, is now a government of the government, by the government and for the government. The people are secondary. Politicians take care of themselves. They live well. They cater to the rich and they don't care about the middle class working people who pay the rich and support the poor. The rich are paid millions of dollars annually. Money to pay the politicians, celebrities, sports figures, corporate CEOs and stock holders, comes from the middle class people. The middle class also supports the people on welfare. Greedy corporations and laws passed by greedy politicians are wiping out the middle class people. When they are gone, everything will collapse.

Before there were income taxes, we had streets, sidewalks, highways, utility companies, streetlights, police departments, fire departments, defense systems, courts, and politicians. Money was even available to finance wars. Now, even with high taxes, we receive little more. And citizens have little say when it comes to taxes, licenses, fees and outrageous penalties that are imposed upon them.

I love America and I know I am lucky to be living here, but that doesn't mean everything done by our government is right.

Something that I found to be quite interesting is that a friend of mine was elected to a very high state office. He went to Sacramento with intentions of rectifying some of the problems with our state government.

"As soon as I tried to make changes, the incumbents with power instantly put the screws to me," he said. "It's like the Devil is running the government."

Sadly, that's the way it is, not only with our government but all the governments in the world. Greed, drugs and the abuse of power create the problems we are forced to live with. If those in power cared and acted accordingly, it could all change.

CHAPTER 78

Greed has destroyed almost every downtown-shopping district from coast to coast. Until the 1960's, every city in the nation—large or small—had flourishing downtown business districts. Local businessmen owned most of the stores. Their businesses thrived, people were employed, and the money spent at the stores stayed within the community.

In the late 1960's, big businesses and malls started their war on the small businesses, wiping out locally owned stores and downtown shopping districts. The sad part is that the city councils of the cities took the tax dollars paid by the small businesses and gave them to big businesses, along with other incentives, to move into the towns and wipe out the little guys. As a result, almost every downtown area from coast to coast has been destroyed.

Business districts once filled with the hustle and bustle of people spending their locally earned dollars are now deserted. Buildings that were once beautiful and well maintained are vacated and crumbling. Many districts have become filthy havens for dysfunctional people, drug dealers, and criminals, and it's no longer safe for anyone to walk those streets alone.

Until the advent of malls and the invasion of chain stores owned by huge corporations, cities and the quality of life constantly improved. Now, thanks to greedy rich people, corporations and governments, many of our cities have been destroyed and the quality of life for those living within them will never be the same.

Unfortunately, the heads of large corporations, like many other people, have become totally controlled by money and material values. They pay themselves outlandish wages, yet they cringe if a worker on the bottom of the totem pole wants a twenty-five cent an hour raise.

Prices wouldn't be as high if corporate heads were paid a fair wage. But they rape the corporations and are paid millions per year. No one earns that kind of money. One hour of their lives is not worth tens of thousands of dollars more per hour then the lives of people who work at lesser jobs.

Talented or highly educated people should be fairly compensated. But they don't deserve the millions they are overpaid.

If all people received fair wages, the people on the bottom wouldn't be so poor and prices wouldn't be so high. The people on the bottom of the ladder buy the products or tickets that pay the people on the top. Many people can no longer afford to purchase anything that isn't a necessity.

My hometown is an example of what greed has done to most communities in our country. About 12,000 people live in Taylorville, Illinois. It is the county seat of Christian County. I was fortunate to have lived there during the forties and fifties, when Taylorville was a flourishing community.

A stately county courthouse with a four-sided clock tower still sits in the center of Taylorville's downtown square. When I lived in Taylorville, thriving stores operated by local merchants lined all four streets facing the courthouse.

From each corner of the square to the next block in each direction, more stores lined both sides of the streets. Theaters were on the east and north sides of the square and a third was on a northeast side street. The retail buildings were filled with locally owned clothing stores, drug stores, variety shops, shoe stores, jewelry stores, hardware stores, restaurants, etc. Even a candy store with a soda fountain was located on one corner of the square. It was always the favorite hangout of the high school kids, and when I was a teenager it was my home away from home.

From the time I moved to Taylorville at the age of four, I looked forward to going downtown. In the daytime, people were shopping, the town was alive, and it was hard to find a place to park a car.

At nighttime, show windows and signs were lit up and the grand marquees on all three theaters glittered with flashing lights and brilliantly colored neon. Only a Las Vegas casino could have rivaled their beauty. Taylorville was at its finest. Growing up and living there couldn't have been better.

During the last few decades things have gradually fallen apart in Taylorville. The marquees have been removed from all three theaters and two of them have closed. There is still one theater on the north side of the square, but the flashing lights and neon are gone. A strip center was built a few blocks from the downtown square, but it didn't have a detrimental impact on all local merchants.

In the late 1960's, chain stores and malls started to invade bigger cities. They destroyed the old downtown business districts in every city near them. Now corporate-owned malls thrive from coast to coast, and the same big corporate-owned stores are in most of them.

It wasn't feasible to build malls in small towns like Taylorville. So business districts in those towns managed to survive—that is, until huge cut-rate stores like Wal-Mart invaded them.

Five of the richest people in the United States are the owners of Wal-Mart. They are worth billions of dollars each. Their money was earned at the expense of low-paid employees, foreign labor, and the small businesses and towns they have destroyed. Many towns like Taylorville are now filled with vacant buildings, or tenants who barely survive by operating antique shops and second-hand stores.

Most of the clothing stores, hardware stores, and department stores that once flourished in Taylorville's business district are gone. Only a couple of the original drug stores, several small shops, and bars remain.

My parents are also gone, but I still love to go back to Taylorville and see the rest of my family and friends. Every time I'm there, I park the car somewhere on the square by the old courthouse. I listen to the bell ring in the clock tower, and I look at the old storefronts of the buildings. Once again I see the beauty of the flashing lights and brilliantly colored neon of the theater marquees, and the hustle and bustle of people shopping with local merchants. I hear Elvis, Connie Francis, and the Platters, and I join my high school friends for another soda in the candy shop, as I reminisce about those wonderful happy times in the '50s that I was lucky to enjoy.

Unfortunately, the quality of living will probably never again be as good as it was during the 1950's. Today's generation will never know what they have missed. Our leaders have not created a better world for us to live in.

The demise of Taylorville's business district isn't unique. Greedy corporations and corporate-owned chain stores have destroyed thousands of businesses and business districts from coast to coast. More people and cities will become their victims unless laws to control greed are placed on conglomerates.

Dollars spent at big businesses do not stay in the local trade area. They are shipped to corporate headquarters to fill pockets of the rich. Dollars spent at small businesses stay in the local trade area and circulate from one business to another. For example: If one hundred thousand dollars is spent at a chain store, the city and state collect taxes on those dollars once, and the balance of the money is sent to corporate headquarters. If the same one hundred thousand dollars is spent at several locally owned small businesses, after taxes the profits usually re-circulate in the local trade area. As a result, those dollars are taxed again and more jobs are created.

Most big corporate heads make more money than they can ever spend, but they become possessed by greed. They aren't appeased until they wipe out all of their competition. It appears to me that Wal-Mart wants to wipe out every business and eventually own it all. It will be a miracle if any small businesses survive.

Big corporations are now victimizing owner-operated tuxedo stores. By the time you read this book, my stores will probably be liquidated. Robinson May Co. bought out a national chain of bridal stores and acquired a nation-wide chain of tuxedo stores. They try to lock every bride into renting wedding tuxedos from their stores. A national chain of men's stores has also attacked the tuxedo rental business.

Unlike locally owned stores, chain stores order their tuxedos from warehouses. Some of those warehouses are thousands of miles away. But the big chains spend millions on advertising to make the consumers think they are getting a better deal, and the small business owner can't compete. When the locally owned tuxedo shops go by the wayside, the customers will lose. Chain stores cannot offer same-day service. There will also be no place left to try a tuxedo on when it is ordered, and some weddings will be ruined because chain stores cannot correct all last minute errors

We do not live in a true democracy. There are no longer equal rights and opportunities for everyone. We live in a capitalist society controlled by corporations and the rich. Big money creates a legal monopoly and unfair competition.

There was a time when conglomerates were forced by law to break up. Monopolies and unfair competition were illegal. But our government did away with laws that protected smaller businesses. Now it's a dog eat dog society and only the rich will survive.

As long as wholesale prices are higher for a small business than for big businesses (because big money buys discounts), the small businesses can't compete.

If wholesale prices were the same for any business, retail prices could still be set at the discretion of the seller. Then, corporate-owned chains couldn't unfairly destroy small businesses and with them our cities. They could no longer control our country or our economy.

Uncontrolled capitalism gives power to greedy rich individuals and corporations. If left unchecked, the middle class will not survive and a society of rich and poor will exist. Home ownership will no longer be in the grasp of working-class people. Like commercial real estate now, homes will be owned by, and rented from, corporations and the rich. "The American Dream" will come to an end.

Capitalism uncontrolled by laws or rules is moronic. It would be wonderful if only honest people who cared for others existed on this planet. But that isn't the case.

We must abide by laws, rules, codes and statutes when we do almost anything in this country. Laws and rules exist because people won't conform without them. Life could not be harmonious. A society devoid of laws and rules would be chaotic. Uncontrolled capitalism—a dog-eat-dog society—is just as dangerous.

Rules to control greed would not end "Free Enterprise." It's foolish to assume that everyone associated with free enterprise is perfect and controls aren't needed. The Roman Empire collapsed because of greed. If capitalism remains uncontrolled, our society too will eventually disintegrate.

Credit cards are another downfall for many people. People are inundated by offers from credit card companies. Those in need get sucked into using a card because companies entice people with low fixed-interest rates. Those rates are no more fixed than the tail on a cocker spaniel puppy. If a cardholder is never late making a payment, the issuer can still double, triple, or in some cases quadruple the interest rate. If they find the least blemish on the consumer's credit report, or at their own discretion without reason, the interest rate can be increased. As a result, many people are never able to pay off their card balances. Almost none of their payments are paid on the principal. Now, thanks to lobbyists and payoffs, the cardholders can no longer file bankruptcy against those companies. Instead of credit card companies lowering fees and interest rates, they are raising them. The cardholders are being legally raped and they will have no recourse.

I told a judge how I felt about credit card companies and asked for his opinion. Even though he agreed with my philosophy, he replied, "Laws are based on the rule, 'Let the buyer beware.' The consumer doesn't have to sign the agreement that gives the credit card companies the right to raise the interest rates without cause, even if the rate is outrageous."

"That is true," I replied. "But the law should protect the consumers. There should be restraints on credit card companies."

The exorbitant interest rates and tactics of the companies that issue the cards should be held in check. They are legalized loan sharks.

The Bible says that the time will come when no one will be able to buy, and no one will be able to sell. It is gradually becoming a world of the rich and the poor. Narcissistic selfish people strive to live excessively lavish lives and/or attain power with no regard for the consequences suffered by those that are less fortunate. Tickets to concerts, theatrical performances, entertainment complexes, sporting events and movies have become so expensive that poor people can't afford to go to them. Little businesses are being wiped out by big businesses and big businesses are being wiped out by giant businesses.

Once we have a cashless society and gigantic corporations and the richest people control everything, *"No one will be able to buy and no one will be able to sell."* Prophecy will be fulfilled.

CHAPTER 79

There is nothing wrong with wanting and having material possessions, if they are not your priority. But a rich man is not always one with lots of material possessions and money. When we die all those things mean nothing. Many poor people will die far richer than some millionaires will. When we die, the only things we can take with us are the things we gave away. Our faith in God and our deeds are the most precious riches a man can possess.

Giving and sharing are two virtues that many people lack. They may have more money than they need, but they are selfish and greedy. They lavish themselves with gifts but will never give anything of value to anyone else. They will spend thousands of dollars on art, for example, and put it in a closet. Yet, they will go to a craft fair and complain that $3.00 is too much to spend on an item hand-crafted by a person who is struggling to make a living. Many people will give away an item they no longer want, but if they sold that item they would never give the money to anyone. Their giving stops when money is involved.

There are many people who are slow to reach for their wallets. They will always let others pick up their tabs, but they never reciprocate. Some people accept invitation after invitation to be entertained and dined but never extend invitations to others. Some people get "bent out of shape" if they think they overpaid when they split a restaurant tab with others, fearing they paid a few cents for someone else. And, some people are showered with gifts but seldom or never give to others. All those people are takers.

That brings to mind a situation I encountered when I was discussing giving to others with a celebrity friend of mine.

"I give a lot to charities," my friend said. "I am constantly going to functions to help them raise money."

I knew my friend never gave anything materially or monetarily to those charities.

"Many celebrities consider going to charity functions as giving," I said. "If they donated to the cause, then they gave. But when a limousine picks them up and takes them to the function, and they are wined, dined, plaqued, applauded and praised; and they leave with more than they went with, how did they give anything,"

"But my time is worth money," my friend replied.

"Two hours of your life is worth no more than two hours of mine," I responded. "The people who give to a charity are the ones who make donations or work their buns off and still buy a ticket. When someone is rewarded to make an appearance, they didn't give anything. They exchanged their time for being pampered, honored, awarded and fed."

My friend didn't know quite what to say, but saw my point. By the same token, rich people who give to charities only to receive honor or to buy themselves into society aren't givers either. They are buying praise, recognition, and social status for themselves. A person can only give when they expect or receive nothing in return.

A few years ago I was in Hollywood with a wealthy friend of mine. We were ready to enter a restaurant when a homeless man stopped us and asked for money to buy food. I reached in my pocket and gave the man some money.

"I wouldn't give that man a penny," my friend said. "Most of those people are phony, and I'm not wasting my money on them."

"I didn't waste my money on him," I replied. "If he's phony, the reason I gave him the money didn't change. I didn't lose a thing. If he's phony, he's the one who lost.

I think the following poem I wrote in 1978 has a good message.

<u>Virtues</u>

Wanting to Give and wanting to Share,
Showing you Love and showing you Care,
Having Compassion for people in need,
Having the Patience to plant deep your seed,
Are all virtues from God, to give but yet keep,
For the seed that you sow is the harvest you reap.

As long as there are people who are givers, people who are takers will take advantage of them. Yet, if givers stop giving to others because takers take advantage of them, they are the losers. The Devil wants people to stop giving and sharing with others. He sees that givers get "burned," hoping it will smother their compassion and consume their good will. He wants us to be self-centered and selfish.

I also believe the surge in racism has been instigated by the Devil. Some people blame specific races for causing problems. But there is no such thing as a race that is a problem. The problem is the bad people of every race. The Devil has to cause dissension and get people to blame the people's race, not their character. It's a shame that people can't see what is really happening and put a stop to it.

Sometimes the leaders of different races are the downfall of their followers. If they don't keep them suppressed, they would lose their power over them and they wouldn't be needed. Instead of keeping people down by convincing them they are victims of society, and everyone else owes them, the leaders should be encouraging them to clean up their act, do something for themselves, contribute to society, and make themselves desirable to be around. Their race is not the problem. Their problems lie within.

No race is free of lazy people who have no goals, no class, no respect for themselves or others, and no desire to be clean. When entire areas are taken over by those people, neighborhoods are destroyed and property values plummet. Unfortunately, when that happens, prejudiced people blame the race of the people who moved into the area and destroyed it. Their race was not at fault. Their character was to blame.

If all people of every race took care of their homes and were assets to their community, it would help put an end to some of the problems. People of every race need to police their own people and promote togetherness with others. We are all responsible for our own lives. We are where we are because of our own choices.

People have also become too touchy. No person or circumstance can hurt us unless we allow it. When evil people make derogatory comments about another person, they are the fools. But if we allow our lives to be affected by those comments, we are the fools. Words can't hurt us unless we let them. Unfortunately, many people jump on their bandwagon and look for every reason to be upset. Instead of being strong, they crumble and let themselves become victims of those who are racist or verbally abusive.

Some people allow themselves to become upset if a person of another race, lifestyle, or creed says a word they think is offensive. Yet, they fall out of their chairs laughing if one of their peers says the same thing. That is stupidity. If it is the word that is devastating, who said it should make no difference. However, words said by anyone can't hurt us unless we allow it.

Until people are proud of their race and people of all races can laugh together, racism will never end. For example: "The Beverly Hillbillies" was a comedy based on uneducated white people with no class. They were misfits in high society. It was a funny series and people of all races laughed together. The only difference between "The Beverly Hillbillies" and "Amos 'n' Andy" was the color of the people. Yet touchy people who could find no humor in a satire about people of their own race cried racism. People need to stop looking for reasons to be offended.

People with a good sense of humor deal more easily with problems and suffer less stress. Negative people who look for reasons to be upset will never be at peace.

People can't continue letting the bad people of every race promote racial problems. Some people who accuse others of racism are themselves the most racist. The Devil wants people to hate one another. He doesn't want people to love one another and live together without prejudice. That's why the KKK, Skinheads and other organizations that promote hatred exist.

Whether we are white, black, brown, yellow or red, and no matter what creed we may be, we must strive to stop people who advocate discord with others. We must love one another. We are all God's people, and by working together we can bring an end to racism and bigotry.

The lack of respect for others is another issue with many people. They have short fuses and they look for reasons to get upset. They don't enjoy being compassionate and courteous. They attain great satisfaction from being discourteous and rude. Yet, they become outraged if someone is discourteous or rude to them.

The Devil knows what makes people angry. He will see to it that whatever triggers their outrage always happens to them. For example: Most of us have gone to restaurants with people who always find fault with their food or service. Instead of being courteous and polite and quietly asking their server to make necessary changes, they become enraged and take their anger out on the waiter or waitress. Sometimes they are loud, demanding and obnoxious. They couldn't care less if they ruin the evening for people sitting with them. Those people are empowered by humiliating someone who is defenseless. Employees of businesses usually can't retaliate when a customer is verbally attacking them. Therefore, rude people have no fear of being counterattacked. They know management will probably pacify them and they are usually rewarded with discounts.

The customers are not always right, especially when they become abusive and belligerent. If business management told people with nasty attitudes to leave and never come back, those people wouldn't continue to mistreat others. They do it because they know they will get away with it. You can't stop the Devil by patting him on the back and allowing evil to prevail.

As I said before, the Devil knows what triggers people's anger. Whatever it may be, he sees that it constantly happens to them. And, every time they become enraged, he wins. They make fools of themselves, and he sits on the sidelines and laughs.

Road rage, retail rage, and rage in the workplace are other examples of the control the Devil has over some people. He has even enticed many of them to murder others over things that were purely accidental and unintentional. Instead of getting satisfaction from being polite, understanding, and godly, their fulfillment comes from being outraged and evil. Kindness, patience, and compassion are not a part of those people's makeup.

Included in that same group are people who are filled with jealousy, anger, and hatred. Their satisfaction comes from promoting discord and conflicts. They bad-mouth other people behind their backs, tell people off, hold grudges forever, and never forgive and forget. Yet, if someone disagrees with them or tells them to take a good look at their own faults, they are enraged. They can dish it out, but they can't take it. Instead of being happy and enjoying life, they constantly find fault with someone or something. They usually spend their lives being "back-stabbers," "pot-stirrers," and "mud-slingers." They love to cause problems between other people. They're possessed by a bitterness that poisons their own lives, and as a result they are poisonous to other people. They destroy families and friendships. They thrive on resentment, anger, and hatred.

Many bitter people have disagreements with their parents, siblings or friends. Instead of being peacemakers and making amends, they hold grudges and never forgive or forget. Yet, many of those same people go to church every Sunday and profess to be godly people. Their spirituality or religion is in their mouths, not in their hearts. Instead of feeling guilty for their own evil actions, they are fulfilled, and they place all the blame on other people.

Evil people may be proud of their evil deeds and may get by with them. But in time they will reap their harvest. Sooner or later they must face their Maker.

Another trick of the Devil is to convince people their problems are always someone else's fault. Greedy attorneys and fallacious psychiatrists help promote that evil. Instead of encouraging people to be strong, cope with their problems, and go forward with their lives; they manipulate them into believing they are marked victims and their lives will never be the same. They encourage people to be weak. They must place blame for their problems on someone else, usually someone with a deep pocket. Only outlandish monetary settlements will get rid of their problems.

The laws and the legal system are responsible for much of the degradation of society. People are no longer held accountable for their own actions. For example, there was the case where a lady was burned when she dropped a cup of hot coffee in her lap. Instead of taking responsibility for dropping the coffee, she sued the restaurant (the deep pocket) for making the coffee too hot. She was awarded millions. What a joke! Coffee is supposed to be hot. Customers complain when coffee is cold. She knew the coffee was going to be hot before she touched the cup. She dumped her cup of coffee in her own lap. The restaurant didn't do it.

Frivolous lawsuits reap millions for greedy attorneys and greedy clients. That is legalized robbery. People should not be allowed to sue for outlandish sums of money they never lost or will never lose. Monetary damages should not be awarded when there were no actual or projected monetary losses. Greedy attorneys and clients who use our warped legal system for monetary gain and not for justice are stealing. Many innocent people are sued because they have deep pockets. No one should be responsible for someone else's actions, unless they were aware of those actions and did nothing to curtail them.

Another example happened in my area a few years ago. A female employee of a local school district sued the district for millions because a male employee took her hand and placed it on his crotch. The female employee went to court and read a five-page letter stating her marriage was ruined and she could no longer trust her husband, or have a good relationship with her children, because of the man's actions. Her life had been destroyed, and she held the school district responsible.

I don't understand how our laws allowed a case like hers to end up in court. The school district was not responsible for what the man did, but because of the possibility of being awarded a large settlement, they were sued.

Thirty years ago a woman in the same situation would have slapped the man's face and told him to never do it again. Within a week, after she told everyone else about it, the man would have been publicly ridiculed and she would have been over it. That would have been the end of it. But now, because of greedy attorneys and misguided counselors, some people can't cope with life or deal with problems until they are paid. As soon as they get a settlement, their life goes on, and they're fine.

Annually, tens of thousands of people are robbed, physically attacked, held at gunpoint, and shot. Yet, they cope with the trauma. They get over it and lead normal lives because the perpetrators of the crimes are seldom caught and there is no one to sue. Even if criminals are caught, they seldom pay restitution to their victims, let alone punitive damages.

I was robbed at gunpoint, so I speak from experience. I was unable to sue the criminals who robbed me. I didn't get my jewelry back and I couldn't get money to compensate me for the loss. I had to be strong. I couldn't disintegrate and let it destroy me. I dealt with the trauma, I got over it, and I went on with my life. If I could have made a choice, I would rather have had my hand placed against a man's crotch then have his gun placed against my head.

An attorney would never file a suit for punitive damages against a criminal for showing you that he has a gun, because most criminals have no money. But if a man indicates he has a penis, which I'm not condoning, an attorney will have him in court before he has time to zip up his pants. And he, his employer, or both will be sued for millions. That is, unless he is found to be penisless or penniless.

Defendants who win lawsuits still lose. Many have lost their life savings defending themselves against unfounded or frivolous charges. People who file lawsuits and lose should be held responsible for damages and court costs incurred by the defendants. That would deter people from filing frivolous lawsuits, and it would help bring an end to legalized thievery. Granted, restitution for monetary loses should be paid to victims who have lost or will lose income because of acts committed by another. But punitive damages are out of line. They are allowing clients and attorneys to legally rob people, companies, and as an end result, you.

Out-of-court monetary settlements have kept a lot of wealthy people, who may or may not have been guilty, from going to trial on criminal charges. In return for payoffs to their accusers, all the charges were dropped. They didn't go to trial and they didn't admit wrongdoing. They literally bought themselves out of trouble. That is an opportunity a poor person can't afford. As a result, many of the guilty rich are free to walk the streets while poor people are prosecuted, convicted, and sentenced. It also proves the accusers didn't file charges to see that justice was done. Their concern was for how much money they could get from the defendant.

Overzealous, money-hungry attorneys have destroyed the legal system. Their fees are outrageous and many of them only take cases worth large sums of money. Many cases go to court based on monetary value, not the concern of justice. As a result, many cases involving wrongdoing against the rich end up in court, and cases involving wrongdoing against the poor do not. Most often the poor man can't afford to file a lawsuit, and there isn't enough money involved in his case to entice an attorney to take it on contingency. Most poor people can't afford civil or criminal justice. It is a luxury for the rich.

The rule in America used to be that a defendant was innocent until proven guilty. That is no longer true. An evil person can destroy someone by calling the police and making false accusations. With no proof the accuser is telling the truth, the accused is usually arrested and put on trial. As a result, many innocent defendants are unfairly convicted based solely on the word of their accuser. **"Thou shalt not bear false witness against thy neighbor."**

Many times people have recanted false accusations after the accused was convicted. But in some cases, the courts wouldn't rescind their sentences and the defendants were still fined and imprisoned. Even if a falsely accused defendant isn't convicted, his life is emotionally and financially destroyed. People who falsely accuse others need to take a good look at the evil they have done, ask for forgiveness, and change their ways.

These are a few of the things I think the Devil has influence over in this world. There are many more.

CHAPTER 80

If the Devil's wave of evil continues to spread throughout the world, chaos will rain upon us. Drugs will destroy nations, race wars and religious wars will abound, the poor will wage war against the rich, corporations will crumble, and nations will rise against nations. People will revolt against tyranny, governments will collapse, and innocent people will suffer. Terror will be rampant in all nations and no one will be immune to the evil that will prevail on earth. It is all happening. Only we can stop it.

Ironically, the first week in September 2001, I dreamed I was riding in the back seat of an open top military jeep. Two soldiers were sitting in the front. Fire was everywhere. I held my hands against the sides of my face and screamed.

"Oh my God, I can't believe the United States is being attacked."

At that instant I woke up. I was relieved to realize it was a dream. I dismissed it for what is was and went back to sleep, but the next morning I felt quite disturbed. That dream seemed too real. I couldn't get it out of my mind. I had plans to take a trip the following week and I felt apprehensive about going on a plane. That dream was constantly on my mind. But regretfully, I didn't mention it to anyone.

On September 11, 2001, at approximately 12 a.m., I flew out of Los Angeles International Airport with a group of friends, headed for Paris, France. During that flight I again thought about my dream. I almost mentioned it to my friends. But fearing they would think I was trying to scare them, I didn't. I regret my silence.

Our plane landed safely at Orly Airport in Paris. As soon as we stepped off the plane and into the terminal, several French people approached us.

"Oh, I am sorry. I am so sorry for what happened to America," a lady said.

We didn't know what she meant. We looked at one another inquisitively.

"What's that about?" We asked one another.

We only took a few more steps before someone else made a comment.

"I'm so sorry for America, so many people killed."

Now we knew something was wrong. We wanted to know what they were talking about. Our tour guide got our group together and told us the news.

"The World Trade Center in New York was bombed this morning," she said. "Approximately 50,000 people were killed. The Pentagon was bombed and half of the White House is gone."

When we heard the news, we freaked. We were in Paris and we had no way to get back home. All airports were closed. Our flight was about the last one that was allowed to continue on to Europe. We were beyond New York City and over the ocean when the Trade Center was attacked. All other flights had been diverted to Canada.

We were speechless, and our eyes were filled with tears. We were afraid and could do nothing about it. I immediately thought about my dream. I knew it had been a premonition.

As soon as we checked into our hotel, we went to our rooms and watched BBC News. We were devastated. But we felt some relief when we learned the White House hadn't been hit.

Despite the devastating news, we enjoyed our time in Paris and the French people were wonderful. At the Palace of Versailles, the flag was at half-mast, three minutes of silence was observed across Europe and the American National Anthem was played.

Outside the American Embassy, flowers, candles, banners, flags and notes from the French people were stacked three feet deep. I picked up several of the notes and read them. One was especially touching.

"America saved us in World War II. We love America and we'll support you all the way."

When I read that note I felt great love for America and a deep respect for the French people. With tears in my eyes, I laid the note back on the mound of flowers and walked away.

By day we held our feelings inside and enjoyed our time in Paris, but we were up most of the night watching BBC News. We were afraid more attacks would come before we returned home. Fortunately, our fears didn't come to fruition, and our return flight to Los Angeles was only delayed one day. When the wheels of our plane touched the ground at LAX, we breathed a sigh of relief and applauded.

Since the commencement of my mission, I have said, "A wave of evil is going to spread around the world and the world will never again be as we know it." That wave of evil has come to pass, but I believe we have only seen the beginning. Is it possible Osama Bin Laden is the "False Prophet," — the false religious leader that we read about in the Book of Revelations? Is the Anti-Christ or Armageddon going to be at our doorsteps soon? A wave of evil is spreading throughout the Middle East. The forces of evil are using religion to promote death, hatred and war. If the free countries of the world stay bound by their own morals and allow that evil to spread, the end result will be Armageddon. Evil has no mercy. God will be our only salvation.

In the last scene of my screenplay, written years ago, a television news anchor is reporting evil events and acts that are occurring around the globe. He ends his telecast by saying, "A wave of evil is spreading around the world." A man watching the newscast turns to the lady seated beside him, unaware that she is a Devil's Disciple.

"You know, it's getting pretty scary out there," he comments.

"You haven't seen anything yet, It's going to get a lot worse," she replies with a baneful glare and an unsettling grin.

I hope I'm wrong, but I believe that statement is true.

Terrorists have changed the world we live in. Under their reign of evil, living on earth cannot be the same as it was in the past. America and the coalition have taken the first step toward bringing an end to "The Devil's Reign." They have declared war on terrorism. But when that mission is accomplished or in check, they must take the second step and stop the evil that thrives within their own countries. All people who promote evil must be stopped.

Gangs are domestic terrorists. They thrive within the borders of most countries. They thrive because most governments do very little or nothing to stop them. They are also a threat to the entire world. If they are allowed to flourish, in time they will destroy every country in which they thrive. When citizens have to give up their freedom and live in fear, government is not doing its job. One of the main purposes of any government should be to protect its people from any evil.

Unfortunately, in the United States evil hides behind the First Amendment. That amendment is constantly abused and misused to allow evil to flourish. The ACLU constantly uses it to fight for the right for people to promote evil. Promoting evil should not be protected by the Constitution. The KKK, Skinheads, the Aryan Nation, gangs and like groups that promote hatred and harm to others should not be allowed to flourish in our country. They are as dangerous as any terrorist. Some of those groups even have training camps where they brainwash followers and promote the spread of their evil. Their leaders are no better than Osama Bin Laden and their groups are no better than Al Qaeda. They must be stopped.

The Fifth Amendment to the Constitution also allows evil to flourish. Using the Fifth Amendment—refusing to answer a question because of self-incrimination—is a legal way for a defendant to not tell the truth to the court. An innocent person should have no need to use the Fifth Amendment. The right to safeguard evil should not be protected by the Constitution.

People who lived in a different world adopted our Constitution. The Constitution was written when life was simple, and most people wanted to make the world a better place in which to live.

Once upon a time in most cities it was safe to take a stroll at night. Many people never locked their doors. But times have changed. The world as it was in 1776 no longer exists. Laws should be adapted to relate to the world we live in today.

Because of terrorism, some new laws have been passed and the government has strengthened security. But more laws should be changed and enforced to help fight the spread of gangs, drug dealers, domestic terrorists, and other evils that we are forced to live with. Right is right and wrong is wrong no matter how you look at it. Laws should be based only on what is right or wrong. Politically correct should not exist.

Not only our government but governments of all countries must stop being complacent. Any action that promotes harm to a government of any good country or to its citizens should be illegal and forcibly stopped. Peace on Earth cannot be attained unless all governments fight evil within their own borders and work with other governments to fight evil in the rest of the world.

Evil attacks Good with a total lack of morals, virtues, conscience or rules. The Devil reigns by thrusting his evil upon the world through evil people that are not bound by regulations. As long as governments and people try to destroy evil and are bound

by rules that protect evil doers, good will never win. **THE LAMB NEVER KILLED THE LION. We must fight with the same rules as our enemy or we will never win.**

People have the power to stop the spread of evil. As I've said before, "We are where we are at this point in time because of our own choices." The world in which we live is the result of choices made by all people. Fortunately, many people chose to live good lives and, as a result, many things on earth are good.

At this point in time, the world is far from being entirely bad. However, if more people strive to be godly, the world will become a better place in which to live.

Even if the world exists forever, time for us is short. We are not immortal. But if we make good use of the life we are given, we can help bring peace and happiness to people on earth. No matter what race or creed we may be, we all instinctively know right from wrong, and most of us believe in a Supreme Being. What we call Him doesn't matter. Allah, Jehovah, or God means "A Supreme Being." We must get our lives in order with Him before it's too late.

God gave us Ten Commandments. They are good rules that everyone, regardless of religious beliefs, should want to live by.

I. **"Thou shalt have no other gods before me."** There is only one God. When you need help, He is the One to go to.

II. **"Thou shalt not make unto thee any graven image."** When God gave us this commandment, graven images were common. Now many new things can become idols to us. Anything that comes between you and God is an idol. It could be your spouse, material belongings, money, celebrities, etc. We should always love God more than anything else."

III. **"Thou shalt not take the name of the Lord thy God in vain."** Sadly, God's name is used in vain so much that most people don't even give it a second thought.

IV. **"Remember the Sabbath day, to keep it holy."** Unfortunately, in these times, many people don't take time to worship God and thank Him for their blessings.

V.   **"Honor thy father and thy mother."** If your parents are godly people, you should honor and respect them.

VI.   **"Thou shalt not kill (murder)."** There is a big difference between killing a person and murdering them. If a person kills another by accident or in self-defense, it is not the same as a premeditated murder committed because of anger, jealousy or hatred.

VII. **"Thou shalt not commit adultery."** When you are married, your spouse should be your only sexual partner. However, that doesn't mean that single people have the right to be promiscuous.

VIII. **"Thou shalt not steal."** You don't have the right to take anything that is not yours. Writing bad checks and not making them good, taking things from your workplace, not paying a valid debt and money obtained from unscrupulous lawsuits are also forms of stealing.

IX.   **"Thou shalt not bear false witness against thy neighbor."** We are not to falsely accuse or speak lies against others. We are to always tell the truth.

X.   **"Thou shalt not covet thy neighbor's house, thou shalt not covet thy neighbor's wife, nor thy neighbor's manservant, nor thy neighbor's maidservant, nor his ox, nor his ass, or anything that is thy neighbor's."** We are not to be jealous and envious of what other people have; we are to be happy with the things God has given us.

As I stated earlier, we are born with a conscience, and we instinctively know right from wrong. An Atheist doesn't believe in God, but should know the rest of The Ten Commandments are good moral rules that conscientious people should strive to abide by.

We control our own destiny. As I said earlier, "We are all where we are by our own choices." To have a good life, we must work for it. Things that are pretty, nice, good, positive or godly do not automatically come to us. We must put forth the effort to attain them. Once they are acquired, we must keep working to preserve them. On the other hand, we don't have to work for things that are ugly, bad, negative or evil. They come to us without any effort. And we must work even harder to keep them out of our lives.

People who put forth no effort to bring positive things into their lives, or to eliminate negative things from their lives, have no excuse to cry, "Poor me." The seed they sow is the harvest they reap. For example: We must constantly work to have clean clothes, a well maintained home and a pretty yard. It also takes an effort to look our best, take care of our body, and to be a good person. If we do nothing, our clothes become tattered and dirty, our homes become filthy and dilapidated, flowers die and ugly weeds take over once beautiful yards. Even our bodies become dirty and ugly.

Our social status and our material possessions are insignificant. Whether we are rich or poor, we must constantly struggle to attract and keep positive, pretty, good and godly in our lives. Negative, ugly, bad and evil come automatically, and they consume the lives of lazy, complacent people.

God knows we are sinful human beings, and He doesn't expect us to be perfect. But as we live with our faults, we must strive to bring good things into our lives and strive to be godly. Every godly thought and deed enriches our lives and our souls, promotes peace on earth, and pleases God. They also diminish evil and weaken the Devil's control over the world.

Drugs, gangs, corruption, greed, terrorism and other depravities are like a cancer spreading a wave of evil throughout the world. If we don't retaliate against all evil with all our might, unbound by rules that protect evil people, there will never be peace on Earth. Our lives and those of our descendants will never be the same.

Most governments of the world have waged war against terrorism. But like I said earlier, that is only one step in the battle against evil. Countries must also fight the evils that thrive within their own borders. Drugs, gangs, greed, corruption, crime and other evils are also threats to a future of peace on Earth.

The Devil, Lucifer, The Evil Force, Satan or whatever you may call him has power over the world only because weak human beings succumb to his lies and temptations. Those people give him his power. Without them, his evilness could not flourish.

No matter what happens to our world, our time on Earth is short. Evil may continue to thrive and we may suffer much pain, but we can save our souls. We must get our lives in order with God before it's too late. If we promote Godliness and fight evil, we will take power away from the evil one and give more power to God.

God gave us the power to topple Lucifer from his throne and bring peace to all people on earth. But we must not be complacent. We must fight evil with all our might.

As I have said repeatedly, "We cannot defeat an evil force that attacks with no rules if we fight back restrained by our own moral convictions. You can't beat the Devil by patting him on the back."

The forces of evil are using the name of God to spread a wave of evil around the entire world. The only way to win the battle against that evil is to fight back with no restraints. We cannot continue to be politically correct and passive. If we have to stop terrorism by bombing entire cities, it must be done. Terrorists will not stop slaughtering innocent people until they fear for the lives of their own families. That is what ended World War II.

If we are complacent and fail to control the spread of evil, prophecy will be fulfilled. Armageddon will become a reality. Then, God will destroy evil people and He will bring an end to "**The Devil's Reign.**"

****

Shortly after I finished writing this story I sent query letters to several literary agents. David Hiatt, a literary agent in Enterprise, Oregon, requested a copy of my manuscript. Shortly after he received the copy he responded. He would represent my book if I had it completely re-edited. The manuscript had been edited. But David insisted he knew what had to be done to make the book marketable. With hopes that David would eventually find a publisher for my book, I agreed to pay him for the edit.

I sent David a printed copy of my manuscript and a copy on a disc. A few weeks later, I received a letter from David. A copy of the original letter is included with other documentation at the end of this book.

255

February 27, 2003

Dear George,

I have enclosed a full-corrected manuscript of The Devil's Reign. After I receive a letter indicating your satisfaction with the editing work, we can begin the marketing effort.

I think there is something dark and unusual with this project. I have had numerous computer problems which appeared only when working with your story. The last one completely wiped out the software, and the computer had to be rebuilt from scratch. I hope the devils are behind us.

Best regards,

David Hiatt

In the summer of 2004, I decided to try another avenue. I put together a folder that contained a short outline of my story, copies of notarized letters from witnesses, and documentation to prove my story was true. I sent one of the folders to Larry Stammer, a writer for the Los Angeles Times. I was under the impression that he was the Religious Editor for the newspaper. I thought by chance he would be interested in doing a feature article on my story for the Sunday Calendar Section.

A few weeks later, I received a call from Larry. He was polite but very blunt.

"We aren't interested in your story." He said. "It sounds so far-fetched that no one is going to believe it. Besides, if you are worshiping God and fighting the Devil, we don't want to hear it. If you are fighting God and worshiping the Devil, then we're interested. That's news."

I had a hard time believing Larry was sincere. That evening I called my friend, Debra Devine. Debra is a gorgeous, voluptuous blonde. She is also the daughter of Diana Dors, who was the British version of Jayne Mansfield. I told Debra about my conversation with Larry.

Debra has always been supportive of my story and me. She has spent a lot of her time and effort trying to help me find success. She too, had a hard time believing Larry was sincere. Neither of us could believe a newspaper would have no interest in covering a true story about fighting evil.

"George," Debra said. "I'm going to call Larry myself. I don't think he really understands what you've been through and what you're trying to do. I can't believe he would make a statement like that."

Debra called Larry. To her disappointment, he said the same things to her. "If you are worshiping God and fighting the Devil we don't want to hear it. If you are fighting God and worshiping the Devil, then we're interested. That's news."

Debra and I were dumbfounded by Larry's remarks. But we knew not to waste words. No matter what we may have said, he was not going to change his mind and help us.

Debra and I tried to find help in other places, but a door never opened for us. No one cared about my story or our efforts to fight evil. No one wanted to help.

That's when Debra and I came up with the idea to organize a group and picket the Los Angeles Times. If the Los Angeles Times wouldn't help fight evil, we would peacefully fight the Los Angeles Times.

During the month of September, nine of my Taylorville, Illinois high school classmates flew to California to visit me. Debra and I decided that would be the perfect time to have the demonstration at the Los Angeles times. Friends who have known me for over 50 years would be there to support me.

On September 22, 2004 at 10 o'clock in the morning, 21 sign-carrying demonstrators peacefully paced back and forth on the sidewalk at the front entrance of the Los Angeles Times.

We hoped other forms of the media would come to see what we were doing. But those hopes were soon dampened. They were not interested in publicizing anything against the LA Times.

Debra decided to call Larry Stammer and have him meet us on the sidewalk.

*Top; left to right: Debra, Larry Stammer
and George in front of LA Times
Bottom; left to right: Part of the group of
friends at the LA Times Protest*

"Larry," Debra said. "We have a surprise for you. There is a friendly group at the front entrance that would like to see you."

Within minutes, Larry stepped out the front door of the Times and onto the sidewalk. He was a bit uneasy when he realized Debra and I had brought a group of friends to picket the Los Angeles Times. But when he realized we were not there to have a confrontation with him, he calmed down.

We had no reason to dislike Larry Stammer. He was a professional man doing his job. And he had every right to disbelieve my story. If he and I could have changed places, I probably wouldn't have believed my story either. It is incredible. There are people who think my story is ridiculous. But regardless of what people think or say, the truth must be told.

Debra and I refreshed Larry's memory about my story, and discussed our previous phone conversations.

"Why are you here?" Larry asked.

"We are here to talk about the failure of the media to help people fight evil." I replied. "The works of evil are publicized by the press daily. If the ACLU or Devil worshipers are fighting God, it's in the news. But if people are fighting the Devil no one cares and the media won't help. You can't fight evil if the media refuses to help. I don't care if there are people who think I'm a nut and won't believe me. But some people will believe my story and maybe their lives will be changed for the better."

"I made the decision to not get involved." Larry replied. "This story is too incredulous. I just can't print it. I'm not saying that you aren't telling the truth. I'm sure you believe what you are saying. But I don't believe it."

"Larry, these people came to help George," Debra said. "They are housewives, mothers, grandmothers, grandfathers, professionals, and civic leaders. They are all quality people who have known George for a very long time. They know George is telling the truth. And, some of them were involved in some of the strange incidents that happened to George. What does it take to convince you?"

"I need proof that all these things are true." Larry replied.

Sharon Younker, one of my high school classmates, spoke up.

"George saw one of our classmates floating above his bed in California the same morning that classmate died in Illinois. George called back to Illinois and told us what he had seen, and he wanted to check on LaDonna. It was impossible that George knew she had passed away."

"I believe that could happen." Larry replied. "I've heard of that happening to people before."

Edie Boudreau approached Larry. "I know George is telling the truth," Edie said.

"Tell me," Larry said. "Would you print this story if you were a journalist?"

"I am a journalist," Edie replied. "I've known George for over twenty years and I was involved in one of George's strange incidents. George was interviewed on a television show and I made a copy of his tape so he wouldn't have to use his original. An evil voice that said, "Lies, Lies, Lies," appeared on the copy I made for George. And no one messed with that tape."

"Have you had an expert examine the tape?" Larry asked.

"No." Edie replied. We had no reason to have an expert examine the tape. We knew it was impossible that any one had messed with it. I later realized my VCR picked up a transmission from my neighbor's CB Radio. It was his voice that was recorded over the soundtrack on George's tape."

Larry didn't believe Edie any more than he believed me. Several other friends tried to convince Larry my story was true. But their efforts were futile.

"You need to get an agent and get your book published." Larry remarked.

"I have an agent," I replied. "As a matter of fact, I received a letter from him. He told me he thought there was something dark and unusual about my story. He loaded it into his computer to edit it. His computer acted up only when he was working on my story. The last time, his computer blew up and he had to have it rebuilt from scratch."

"You'd better get a new agent," Larry declared. "He's shining you on. I believe he told you that. But I don't believe that for one minute. He's playing on your emotions. You'd better get rid of him and get a new agent, or self-publish your book."

Larry didn't believe my story but he knew we were not there to cause problems. He knew our intentions were good. Instead of telling us to leave, or ending the conversation and going back to work, he took more time to talk to us. He even found some humor with our presence.

"Where's all the cameras?" he said with a chuckle. "I've been with the Times for over 30 years. Ink runs in my veins. And this is the first time this has happened to me."

"We've never done this before either," I replied as we all laughed.

The time and effort we spent on that little demonstration was not wasted. If we had not gone to the LA Times, we would not have had that conversation with Larry. He still doesn't believe my story. But that doesn't matter. He was honest. He was polite. He listened, and he was a gentleman. He not only gave us his time; he gave me some good advice.

"If you self-publish your book, your story remains intact," Larry said. "An editor can't change it. And no one can waste more of your time or money."

I heeded Larry Stammer's words. David Hiatt had been my agent for almost two years. But he had never given me the status of any of his marketing efforts. I didn't know if he had ever contacted a publisher. After I spoke to Larry, I received a letter from David. I thought that perhaps I had received good news. I was disappointed to find that he only wanted more money to keep my book listed on his web-site.

I had waited almost two years for David to find a publisher. And, I paid him $6,000 to edit my manuscript. He said the editing had to be done before he could market my book. I paid him the money with hopes he would find a publisher. Now, he wanted more money to keep my book listed on his web-site. His letter made me realize it was time to make a change. I had to stop handing people money and waiting for "nothing" to happen. I had to do it on my own.

Thankfully, I followed Larry's advice. I did it on my own. My book was published in August 2005. Unfortunately, the $6,000 I paid David Hiatt for editing work was a waste of money. His editing was not necessary.

Words can't describe how thrilled I was to see the first copy of my book. It was everything I wanted it to be. Even the dust cover, which I had designed, was perfect.

In September of 2005, over one hundred people attended a party to celebrate the publication of my book. My friend Kathy Bee, who is a wonderful singer and hostess of her own television show, donated her time to entertain everyone, and there was a ton of food and drinks. It was an exceptionally exciting day for me. I had fulfilled my mission. The years of trials and tribulation were behind me and I was ready to promote the book I had struggled so hard to complete.

I hired a publicist and started an advertising campaign. There were a few newspaper articles about my book and I was interviewed on several radio and television shows in California, Washington and Kentucky. One of those interviews was on a radio station in Sacramento, CA, the home of the Governor's office. On that show I talked extensively about Robert Munger taking my money, destroying my script, and the fact that over twenty scenes of my script appeared in Arnold Schwarzenegger's film "End of Days."

In October of 2005, I went to Sacramento to visit my former neighbors Rosanna and Jeff. To my surprise, Rosanna had a connection to get me into Arnold's office.

"If you want to take a copy of your book to Arnold, I will take you to his office," Rosanna said.

I never thought I would ever have that opportunity and it only took me a second to respond. "Yes, that would be awesome, maybe it was meant for me to come to Sacramento so I could do that."

The following day, Rosanna and I went to the California State Capitol Building. Arnold was out of town so we placed an envelope containing a copy of my book and a letter to Arnold on his desk in the Governor's office. The letter read as follows.

October 19, 2005

To:
Governor Arnold Schwarzenegger
Sacramento, CA
Re: Book, "The Devil's Reign."

Dear Governor Schwarzenegger,
Please find enclosed a copy of my book, "The Devil's Reign." I have spent over twenty years of my life trying to tell a story that

will change people's lives for the better and help make the world a better place to live.

Since 1998, I have tried to contact you several times about the film, "End of Days." I think I was unsuccessful because my book had to be published first. You must read my book and you will know why I was devastated when I heard "End of Days" was going into production.

I know you are a man of high integrity and honor. I supported you in your run for Governor of California. I also know that you have no idea that "End of Days" was ripped off from me. I believe you will stand up for what is right. I have documentation to prove everything in my book is true. I have original scripts, canceled checks and witnesses to prove Robert Munger ripped me off and stole my story. In one of the scripts, which I paid for, he even changed the name of my main character in the film from Amy to Christine.

I want you to know I am not out to sue anyone. I only ask that you help me so the truth is told. We have mutual friends who will vouch for my honesty and character but they won't say anything until they know you will be willing to listen to the truth. You may investigate me and my story. You will find that my book is true.

I am going on TV and radio interviews and I want to be able to say that you are doing everything you can to support the truth. I hope to hear from you soon.

Thank you,

George Newberry

I thought I would receive a response from Arnold, but it didn't happen. Less than a month after Arnold had the copy of my book, I received a call from my publisher.

"We are severing all ties with you and canceling your book unless you remove everything concerning Arnold Schwarzenegger and his movie, including the name of the production company, from your book."

I couldn't believe what I had heard. "I can't remove it from my book," I replied. "It is the truth and I have documentation to prove it."

That didn't matter. The following day, my book was placed on hold by the publisher.

The Daily Bulletin Newspaper published a third of a page article about the dilemma I was having with my book. A reporter tried numerous times to contact Arnold Schwarzenegger, but he didn't respond. I was disappointed that Arnold had no interest in investigating my story or learning the truth. I can only assume that he didn't care what had happened to me, and he chose to protect his friends.

I called the publisher's representatives and asked them to send me a list of anything they questioned in my book and I would send them documentation to back up my statements. They agreed to do that, but the list never came. Over a year passed by and the problem was not resolved. During that year, I was unable to purchase copies of my book and it was impossible for me to promote it.

My publisher finally agreed to connect me with a New York attorney, at their expense, and furnish him with documentation to validate statements I had made in my book.

The attorney and I had an excellent rapport. We worked well together and I trusted him. He was as concerned about me and my book as he was about protecting my publisher. It was a winning relationship for all concerned. He reviewed my book and I furnished him with all the documentation he requested.

I was elated when I was told my validation was sufficient and my book would not be changed. Authorhouse, my publisher, also gave me permission to update my story and agreed to republish my book.

I felt that the devil had been trying to stop me from telling my story again. Thankfully, if that was the case, he didn't win.

I must say something in support of my publisher. I understand why there was concern about some statements I made in my book. Without documentation, they could have been conceived to be libelous. With mutual cooperation, Authorhouse and I resolved those problems. That is better for my publisher, my book, and me.

Authorhouse is a good publishing company. They have produced an excellent book for me. I recommend them to any author who wants to publish a quality book.

******

I hope the people who read this book are inspired to live better lives. The future of mankind is in our hands. We can remain politically correct and allow the wave of evil that is spreading around the world to continue, or we can fight back with equal force and stop it. Our choices will determine whether we will live in a godly world or continue to live during "**The Devil's Reign.**"

ACKNOWLEDGEMENTS

I give special thanks to my two daughters and their husbands, **Tom and Kathy Unsell** and **Virgil and Vicki Buckner** for being supportive and listening to me during all the years I've cussed and discussed working on this book.

Randy Del Turco – Ontario, CA - My business partner and best friend. Thanks for twenty-five years of encouragement and support. You are one of the few people who watched everything happen and never told me I should give up and quit. You were always there during the bad times and kept encouraging me to follow my dream and never give up. Thanks for walking in the demonstration.

Bob and Cindy Mills – Chino Hills, CA - My close friends for over twenty years. I owe you a special thank you for being supportive and standing beside me through thick and thin. Thanks for believing in me and being there when I needed your help, and for not being afraid to tell people that my story is true. Cindy, thanks for walking with us in the LA Times demonstration.

Debra Devine – Hollywood, CA - I thank you for being supportive. Thanks for helping with the LA Times demonstration.

John Buonomo – North Hollywood / Palm Springs, CA - What can I say? Thirty years of friendship speaks louder than words. Thanks for always being there for me. Thanks for your letter and legal deposition that support the truth.

Orland DeCiccio – Ontario, CA – My neighbor and friend for over 35 years. You've encouraged me and given me your support, loyalty and prayers from the beginning. You stood beside me and believed in me. When I was down, you gave me the strength to carry on. Kudos to you.

Sharon Younker – Taylorville, Illinois – High School classmate and friend for over 50 years. Thanks for your support. We have had many phone conversations and visits since I started working on this book. Thanks for being a friend and for walking in the demonstration. Also, thanks for your letter of support and for verifying my 1976 phone call about LaDonna.

Marilyn Malmberg – Taylorville, Illinois – High School classmate and friend for over 60 years. What can I say about you? Not enough! You've always been a sweetheart and my good friend. Thanks for your letter of support and for verifying my 1976 call to you concerning LaDonna.

Christina Miller – Los Angeles, CA – LaDonna's cousin. The way we met was a mind blower, but it was no accident. Without you I would never have seen the script for "The End of Days." It's a shame that everyone who said they were going to produce "The Devil's Reign" were either fabricating stories or telling lies. I was no more devastated than you were. But that will change. Thanks for always being there for me.

Edie Boudreau – Fontana, CA – Writer and my friend for over 25 years. We met when I first started to work on this story. You came into my life to help me and you've been a part of my life ever since. Thanks for your support through these years of trials and tribulations, for you letter validating my statements, for walking in the demonstration, and for all the time you spent editing my book.

Tom Hubbell – Woodland Hills, CA – Writer and friend. Thanks for believing in my story and my mission enough to walk away from Bob Munger. Two good things did come from my meeting Bob, my friendship with you and Jim Hardiman. Thanks for coming to my aid after you realized Bob was taking my money and had no concern for my script. You have been a good friend.

Jim Hardiman – Cathedral City, CA – Writer and friend. Thanks for your efforts to save my story. You are a very spiritual man. I am fortunate to know you. Thanks for your letter validating statements I made in my book concerning our relationship with Bob Munger.

Jerry and Rosemary Barnes – Palm Desert, CA – My former neighbors, and my friends for over 15 years. Thanks for your friendship and for encouraging me to not give up and quit. Rosemary, thanks for your legal deposition verifying the truthfulness of statements I made in this book.

A SPECIAL THANKS

Writing this book hasn't been an easy task. The support and encouragement of my family and friends kept me working on this book and helped me fulfill my mission. I don't know if I would have kept going without them. My high school classmates flew to California to visit with me and joined my local friends to support me in the Los Angeles Times demonstration.

John Rillo – Friend for 20 years. Claremont, CA

Jim Medlin – Friend for 25 years. Upland, CA

Jayne Stephenson – Friend for 30 years. Murrells Inlet, SC

Terry Dunn – Friend for 35 years. Colton, CA

Mary Bell – High School classmate. Friend for 55 years. Taylorville, IL

Hazel Dobyns, High School classmate. Friend for 55 years. Pensacola, FL

Janet Eton – High School classmate. Friend for 50 years. Peoria, IL

Rob Coatney – Los Angeles, CA Friend.

Rudy Tronto – Studio City, CA Friend.

Ron Anderegg – Los Angeles, CA Friend

Alice Montgomery – High School classmate. Friend for 55 years. Washington, PA

John Olietti – Palm Springs, CA. Friend

Emily Stringer – High School classmate. Friend for 50 years. Edinburg, IL

Marilyn Voggetzer – My first grade girlfriend. Friend for 60 years. Taylorville, IL

Del Funk – Palm Springs, CA. Friend

Cindy Newberry – My cousin. Fontana, CA

Patricia Almazan – Friend for 25 years. Upland, CA

Chuck and Jean Ramsey – My sister and brother-in-law. Decatur, IL

Don and Mary Stickel – My sister and brother-in-law. Taylorville, IL

Tom and Carla Murphy – My brother and sister-in-law. Charleston, IL

Ed and Joyce Murphy – My brother and sister-in-law. Rockford, IL

Mike and Cherese McGaughey – Goddaughter and husband. Longview, TX

Rita Reber – Friend for 60 years. Palm Harbor, FL

Thomas and Edith Murphy – My parents.   Heaven

DOCUMENTATION

On the following pages are legal documentation and signed notarized letters of validation from people who witnessed or were involved in the series of strange events that are the backbone of this story.

SEARCH WARRANT

On January 3, 1992, I was working on this story when the doorbell rang. I opened the door and was shocked to see six or eight Sheriff's officers charge into my home. I was handed a copy of the search warrant that is reproduced on the following pages. The officers were looking for drugs. They were on the trail of my cousin and her boyfriend. I told them the results of my efforts to help my cousin. The Ontario Police Department verified my story and the search ended.

When I read the warrant, I was astonished. I only had two cars, a 1981 Buick Rivera and a 1988 Dodge Caravan. The third car listed on the warrant, a **1972 Ford, license plate LUCIFUR**, was not and had never been mine. I called the deputy in charge of the case. He said the DMV sent them a report of all vehicles registered to me. They couldn't explain how the **1972 Ford** appeared on my DMV report.

The police later ran a check on the license plate, **LUCIFUR**. The car had been registered to another George Newberry in **1972.** Title to the car was transferred to someone in Compton, CA in **1982.**

In 1993, after being the victim of an armed robbery, I realized every strange thing had happened to me for a reason. I was living the story I was to tell. The warrant was now going to be a part of my story. *"Is there a message to this 1972 Ford with the LUCIFUR license plate that I am missing?"* I thought.

I had never done numerology but I knew how it worked, so I decided to try it on the **1972 Ford** and the **LUCIFUR** plate. The results of my efforts were astonishing. Six different columns of numbers each totaled **1998.** I didn't know the significance of **1998** until an inaudible voice said, *"1998 will be the third time something* *will occur since Christ died."* I divided **3** into **1998. 1998 ÷ 3 = 666.** I later discovered I had systematically picked the numbers for each column.

Lucifer is spelled wrong on the warrant, and in error the Judge dated the warrant, January 3, **1991**. The actual date was January 3, 1992. The year had just changed.

If the car had not been a **1972 Ford,** if **LUCIFUR** had been spelled right, if the title transfer date had been any year but **1982,** or if the Judge had written the right date on the warrant, the results of the numerology would not be significant.

On the following three pages are copies of the search warrant and the numerological configurations.

COUNTY OF SAN BERNARDINO, STATE OF CALIFORNIA

SEARCH WARRANT
(PENAL CODE 1529)

THE PEOPLE OF THE STATE OF CALIFORNIA: To any Sheriff, Constable, Peace Officer or Policeman in the County of San Bernardino:

Proof, by Affidavit, having been this day made before me by Richard S. Hahn, a duly authorized Deputy Sheriff for the County of San Bernardino.

THAT THERE IS PROBABLE CAUSE FOR BELIEVING THAT: At the residence located at 303 West Armsley Square, City of Ontario, County of San Bernardino, State of California, there is a quantity of methamphetamine being manufactured for sale, tending to show that a felony is being committed in the County of San Bernardino, State of California, to wit: Manufacturing of Methamphetamine, in violation of Section 11379.6 of the Health and Safety Code.

YOU ARE COMMANDED at any time of the day to make immediate search of the premises located at:

███ West ████████ ███████
City of Ontario
County of San Bernardino
State of California

The residence is described as being a two-story residence with white stucco exterior and having a red tile roof. The front door of the residence faces north and is black in color. There is a detached garage at the southeast corner of the property adjacent to ████████████. The residence is located at the southwest corner of █████████████ and ████████████. The numbers ██████ are painted white on black on the curb directly in front of the residence. In addition, the numbers ████████ are in black metal at the front of the residence.

And all rooms, attics, basements, cellars, safes, vaults, trash receptacles and other parts therein, surrounding grounds, garages, storage rooms and outbuildings of any kind located thereon.

And all persons present during the service of the Search Warrant, their vehicles in which they are in control of as well as all incoming phone calls into the residence relating to illegal narcotic activities.

PERSON TO BE SEARCHED:

NEWBERRY, George Albert
White male adult, 51 years old
Date of birth: November 21, 1940
5'10," 155 pounds
Brown hair, hazel eyes

VEHICLES TO BE SEARCHED:

1. Black 1981 Buick
   California license: 1CRE851

2. 1988 Dodge
   California license: 2XNB997

3. 1972 Ford
   California license: LUCIFUR

FOR THE FOLLOWING PROPERTY:

Methamphetamine, Amphetamine Sulfate, Phencyclidine (PCP), Methaqualone, paraphernalia including lab equipment capable of being used to manufacture the above; together with books and formulas for the manufacturing of these controlled substances, pill labeling machines, plastic bags, scales; together with personal property tending to show dominion and control of the premises and not limited to: keys, canceled mail envelopes, rental agreements, receipts, bills for telephone and utility service, photographs and notices from governmental agencies, to include papers tending to establish a conspiracy, including ledger account books, canceled checks, bank book, correspondence agreements, contracts, phone lists, bills, receipts and record of ownership to vehicles and real property. Also including some or all of the chemicals used to manufacture Methamphetamine (1), Amphetamine Sulfate (2), Phencyclidine (3) and Methaqualone (4), their compounds, salts and mixtures thereof:

Also recordings and notes of any incoming phone calls that might be related with the sales of a controlled substance or narcotic transactions.

Also to be searched for are any amounts of United States currency or other assets or items of value which can be shown to be the result of or involved in the trafficking of a controlled substance.

Ether 1-2-3-4	Hydrochloric Acid 1-4-3
Phenylacetone (P2P) 1-2	Methanol 2-1
Hydrogen 1-2	Sodium Acetate 1
Palladium Black 1-2	Potassium Hydroxide 1-2
Potassium Hydroxide 1-2	Lithium Aluminum Hydride 1-2
Sodium Hydroxide 4-3	Potassium Carbonate 4-3
Ammonium Hydroxide 4-3	Methylamine 1
Ephedrine 1	Hydrogen Iodide 1
Zinc or tin foil 1	Hydroxyl Amine 2
Sulfuric Acid 2	Formamide 2
Piperdine or its salts 4	Cyclohexanone 3
Bromobenzene 3	Petroleum Ether 3
Magnesium Metal Turnings 3	Phenylmagnesium Bromide 3
Isooctane 3	Hydrobromic Acid 3
Sodium Carbonate 3	Anhydrous Hydrogen Chloride 3
Potassium Cyanide 3	1-Piperdine Cyclohexane Carbonitrile 3
Sodium Bisulfite 3	Benzene 3
P-Toluenesulfonic Acid 3	Ammonium Chloride
Toluene 3	Methyl-Benzexazone 4
Toluidine or its salts 4	Acetylanthranilic Acid and its salts 4
Ethanol 4	Phosphorous Oxychloride 4
Chloroform 4	Red Phosphorous

SEARCH WARRANT
PAGE 3

And order the hazardous chemicals and apparatuses be destroyed after the appropriate samples and photographs have been taken.

GIVEN UNDER MY HAND, and dated: *Jan 5, 1991*

Judge of the Municipal Court District
West Valley Division
County of San Bernardino
State of California

1972 + 6694 (Ford) = 8666. ACLU = 6668

1972 FORD LUCIFUR - NUMEROLOGY

1972	**FORD**	**LUCIFUR**
1+9+7+2= 19	6694	3339639
1+9=10	6+6+9+4=25	3+3+3+9+6+3+9=36
1+0=1	2+5=7	3+6=9

1991	1982 — this transfer date was not on
1+9+9+1=20	the warrant. I left it as is.
2+0=2	

```
  1972              1972 (YEARS BEFORE 1982      1972 (
 + 19 (AGE OF CAR        TRANSFER DATE)         + 25 (FORD)
        IN 1991)   + 7 (FORD)                   +20 (WARRANT #
 + 7 (FORD)        + 9 (LUCIFUR)                       DATE
 ─────             ─────                         ─────
  1998              1998                          2017
                                                  - 9 (LUCIFUR)
                                                  - 7 (FORD)
                                                  - 1 (1972)
                                                  - 2 (1991)
                                                  ─────
                                                   1998
```

```
  1972                                          
 + 36 (LUCIFUR)     1991 (WARRANT DATE)         1982 (TRANSFER DATE)
 ─────             + 7 (FORD)                   + 7 (FORD)
  2008              ─────                       + 9 (LUCIFUR)
  - 7 (FORD)         1998                        ─────
  - 1 (1972)                                      1998
  - 2 (1991)
 ─────
  1998
```

```
      666            1972
 3/1998           + 6694 (FORD)
    18             ─────
    19              8666
    18
    18
```

ACLU

I believe the ACLU is the right arm of the Devil. They do just enough good to convince people that they are fighting for our benefit. But they fight for and protect criminals and ungodly people. I believe they are helping promote the spread of evil.

Numerology for ACLU

```
1   13  5  18 9  3  1  14
A   M   E  R  I  C  A  N
1   4   5  9  9  3  1  5   =  37  =  10  =  (1)
```

```
3   9  22 9  12
C   I  V  I  L
3   9  4  9  3   =          28  =  10  =  (1)
```

```
12  9  2  5  18 20 9  5  19
L   I  B  E  R  T  I  E  S
3   9  2  5  9  2  9  5  10
                        1  =         45  =  (9)
```

```
21  14 9  15 14
U   N  I  O  N
3   5  9  6  5   =          28  =  10  =  (1)
```

$$(1) + (1) + (9) + (1) = 12 \qquad 1+2 = (3)$$

```
1   3  12 21
A   C  L  U              ACLU = (1333)  →  1333÷2= 6665
(1   3  3  3)
                                            6665+3=(6668)
```

```
                    6  15 18 4
From Search Warrant 1972 + F  O  R  D
                    1972 + 6  6  9  4  = (8666)
```

275

NOTARIZED LETTERS OF DOCUMENTATION

March 1, 2004

Christina Miller
~~~~ W. ~~ Place Unit ~
Los Angeles, CA 90045
(310)

Re: Verification for George Newberry's Story

To Whom It May Concern:

The week of July 4th, 1998, the hostess of a Raleigh N.C. television talk show, Dea Martin, came to visit with me in my Hollywood home. Dea called George Newberry and told him she wanted to see him before she returned to Raleigh. I had never heard of George and I knew nothing about him. George invited Dea to a barbecue. Dea asked George if it was ok to bring her son, Dakota and me. George said, "Of course."

On July 2nd, we arrived at George's home in Ontario. Dea introduced me to George and several of George's friends. One of his friends present was Wendy Moss, the daughter of Jane Withers (Actress and Josephine The Plumber on the long running Comet commercials.) Dea asked George if he would mind telling me about his movie script. At that time I was working in the entertainment industry. Before George told me anything, I asked him, "Were you raised in California?" George replied, "No, I was raised in Illinois." "What town in Illinois?" I asked. "A small town near Springfield," George responded. "What small town?" I asked. "Taylorville," George answered. I couldn't believe George was from Taylorville. I had been there, and I had relatives living there. Did you ever know anyone there named LaDonna? I asked George. George was startled. "Oh, my gosh!" he replied. "I was getting ready to tell you about a LaDonna I knew, who lived there, but the LaDonna I'm talking about died twenty two years ago." "The LaDonna I knew died young," I replied. "What's the husband's name of the LaDonna you knew?" "Julio Monge" George answered. "That's the same LaDonna," I told George. "She was my cousin."

At that point, everyone in the room was flabbergasted and shocked. They were covered with Goosebumps and exclaiming to one another, "I can't believe this. I can't believe this. This is unreal."

Everyone there knew the story about LaDonna floating over George's bed, except for me. I didn't believe George until he showed me written proof that we were talking about the same LaDonna. Then, I too was in shock. We were all so amazed by what had happened, involving all of us, that we stayed at George's home for four days.

During those four days, we watched the movies The Miracle of Our Lady of Fatima and The Song of Bernadette. In both films, Mary is constantly referred to as, "The Lady." All of a sudden, George put his hands to his face and exclaimed, "Oh! My gosh. I can't believe this. I can't believe after twenty-two years, I just now realized the Italian word for, "The Lady" is, "LaDonna."

George and I instantly became close friends. In 1998, when he read a newspaper article about Arnold Schwarzenegger's film, End of Days. George called me. He was very distraught. The article described the plot of the film. It sounded strangely close to the plot of George's script. I made a call to a friend of mine. The next day George and had a copy of the script for End of Days. We were both shocked when we read it. Too many things in The End of Days script were very close, or identical, to things in George's script. We were also baffled to see that the Production Company was Lucifilms. And, when the production start date was set for November 21, 1998, we were blown away.

George has gone through hell trying to tell his story. He is a good person and he is very honest. His story is true. It is a spiritual story that will change people's lives for the better. I'm sure George and everyone involved with his story will be willing to submit to a Polygraph test to validate their truthfulness.

Sincerely,

Christina Miller

Christina Miller

TIMOTHY MARQUEZ
Commission # 1266034
Notary Public - California
Los Angeles County
My Comm. Expires Dec 2, 2004

State of CALIFORNIA County of LOS ANGELES
On 3/5/04 Before me TIMOTHY MARQUEZ Notary Public, personally appeared CHRISTINA MILLER .

proved to me on the basis of satisfactory evidence to be the person whose name is subscribed to the within instrument and acknowledged to me that she executed the same in her authorized capacity, and that by her signature on the instrument the person or the entity upon behalf of which the person acted executed the instrument.

Witnessed my hand and official seal.

Notary Public

February 23, 2004

Orland DeCiccio
W. ████████
Ontario, CA 91762
Phone (909) ████████

Re: Documentation to validate George Newberry's Story

To Whom It May Concern:

My name is Orland DeCiccio. I have known George Newberry for 35 years. I am one of many people who will vouch for George's character and truthfulness. I met George before he had the dream that started his story. It is a godly story written for people of all races and creeds. It will change people's lives for the better. I have witnessed many of the mind-boggling supernatural events that have happened to George as a result of his efforts to touch people's lives and expose the forces of evil. I have watched him suffer one difficulty after another. A force that didn't want his story written or told was attacking him. Many of George's friends were afraid, and advised George to stop working on his story and burn it. But, George's faith in his mission strengthened and he refused to quit. He knew he was on a godly mission.

Many of George's experiences are unbelievable. And, as much as I have believed in George's credibility, even I questioned his truthfulness about some of the unbelievable things he told me. I soon found out from witnesses and people involved that George had told me the truth. Even they were in complete dismay that such things could really happen. Unlike many people who have had spiritual supernatural experiences, George has documentation and witnesses to verify that he is telling the truth.

In 1991, I was witness to one of George's unbelievable experiences. George's car was broken down and the mechanic couldn't find the cause of the problem. His van was out of commission because his cousin's boyfriend had wrecked it. George had no means of transportation. He drove a rental car for several weeks, then my wife and I came to his aid and loaned him our pickup truck. When the mechanic called George to tell him his car was finally fixed and the faulty part number was 666, I went with George to the auto parts store to verify that the mechanic had told the truth. We thought the mechanic was just kidding. To our surprise, the part number of the faulty new plugs was 666 and the number of the faulty new wires was 666-000-666. You can check an auto supply catalog for the spark plugs and wires for a 1981 Buick Rivera and find this to be the truth.

George's story is true. It needs to be told.

Sincerely,

Orland DeCiccio

Orland DeCiccio

DAVID RODRIGUEZ JR.
Comm. # 1398453
NOTARY PUBLIC - CALIFORNIA
San Bernardino County
My Comm. Expires Feb. 4, 2007

State of CALIFORNIA   County of SAN BERNARDINO
On 2-23-04   Before me DAVID RODRIGUEZ JR.   Notary Public,
personally appeared ORLAND DECICCIO.

He proved to me on the basis of satisfactory evidence to be the person whose name is subscribed to the within instrument and acknowledged to me that he executed the same in his authorized capacity, and that by his signature on the instrument the person or the entity upon behalf of which the person acted executed the instrument.
Witnessed my hand and official seal.

david rodriguez jr.
Notary Public

277

Mrs. Robert (Cindy) Mills
▓▓▓ ▓▓▓ Court
Chino Hills, CA 91709
(909) ▓▓▓-▓▓▓▓

February 25, 2004

Character Reference and Validation for George Newberry's Story

Approximately twenty years ago, my husband and I met George Newberry at a mutual friend's party. When we first met George, he told us the story about seeing his high school friend floating above his bed and finding out she had passed away the same day. He also told us about the near-death experience and dream that prompted him to write a spiritual story. At that time we did not know George well. We recently told George, "When you first told us some of the strange things that have happened to you, we thought you were a nut."

After we got to know George, we changed our minds. We have witnessed George battle a force that doesn't want his story told. At one time, when I had George's script in my home, even I felt threatened. I told George I was afraid and that he should stop working on his story. George told me, "That's what the devil was wants me to do." George knew he could never quit. He knew he was being attacked because his story was against evil and promoted good. He was determined not to be afraid. To him, it was proof he had to accomplish his mission.

In August of 1997, we attended an eightieth birthday party for George's mother. Robert Munger and his wife were present at that party. Bob, as everyone called him, was introduced to everyone as the producer who was going to make George's film. I immediately felt strange vibes from Bob Munger, and after the party my husband and I told George not to trust him. As George has stated, Robert Munger ripped him off and destroyed his script.

In early 1998, on his way home from a Hollywood function, George called me from his cell phone. He was upset because he had told someone he didn't know was a screenwriter the plot of his script. George told me that after he finished telling his story he asked the man what he did. He said the man replied, "I am a screenwriter and I'm working on a similar story." He told George his name is Andrew Marlowe. He also told George that he had written the Air Force One screenplay. George was concerned. He told me he almost asked Andrew if he knew Robert Munger, but didn't. George felt that Andrew was a successful writer and didn't have to steal someone else's story. We were all shocked to learn that Andrew Marlowe wrote the script for End of Days. When George read the script, he told me Andrew had to have had contact with Robert Munger. There were details and scenes in the End of Days script that George had not mentioned to Andrew. They came from a draft of George's script that was only in the possession of Robert Munger.

My husband and I are also witnesses to another scam pulled on George. Daniel DaValle (Danny), who we also know personally, had an agreement to buy an apartment building from George. Danny collected the rent money and didn't make any payments. He lied to George, to us, and to the Mortgage Company. As a result, the building was sold at a foreclosure sale one month before George knew it. George had owned the building for over twenty years.

George is not a religious fanatic. He has a good personality and a wonderful sense of humor. He is nice to everyone and he is extremely honest. His story is true and it should be told.

Sincerely,

Cindy Mills

State of _California_   County of _San Bernardino_
On _March 3, 2004_ Before me _Erica Machado_ Notary Public,
personally appeared _Cindy Mills_

She proved to me on the basis of satisfactory evidence to be the person whose name is subscribed to the within instrument and acknowledged to me that she executed the same in her authorized capacity, and that by her signature on the instrument the person or the entity upon behalf of which the person acted executed the instrument.

Witnessed my hand and official seal.

Erica Machado
Notary Public

March 3, 2004

John Buonomo
████ N. ██████
Palm Springs, CA 92262
(818) ███-████

VALIDATION FOR GEORGE NEWBERRY'S STORY

To Whom It May Concern:

My name is John Buonomo. Jane Withers introduced me to George Newberry almost thirty years ago. I knew George before he had the dream that started his story. I have witnessed the unbelievable chain of events that have occurred to George. I know George very well and I know his story is true.

In August of 1997, I went to the eightieth birthday party George had for his mother. I met Robert Munger at that party. He was introduced to everyone as the man who was going to produce George's movie. I also met Tom Hubbell at that party. Robert Munger had engaged Tom to rewrite and polish George's script. Tom brought a copy of work he had just completed on the script, and gave it to George.

In October of 1997, I was once again at George's home. While I was there, Robert Munger and George returned from Cathedral City with a copy of a script written for George by Jim Hardiman. I knew something was wrong. Robert asked George for money to pay Jim, but George refused to give Robert the full amount due. After Robert, or Bob as he is called, left for home, George told me what was wrong.

George explained that he had been happy with the work Tom Hubbell had done with his script. But for reasons unknown, Tom had stopped all communication with Bob. Bob's inability to contact Tom forced him to find another writer to polish George's script. He had taken it to Jim Hardiman for completion. The script George and Bob brought back from Cathedral City was the results of Jim's efforts. George had read the script when he was at Jim's home with Bob. George was very upset. His story had been destroyed. I read the script and I too read nothing that resembled George's original story. George told me Jim Hardiman was not at fault. Bob Munger had not told Jim he was to polish and rewrite an existing script. Jim did not know George's script or treatment existed. Bob had told Jim he had carte blanche to write a story about a girl who was being pursued by the devil. George was not upset with Jim Hardiman. He knew Bob was to blame.

The next morning I went downstairs to find George very distraught. He had been up all night. He was devastated that Bob had destroyed his story. George called Jim several times in my presence. George was going through emotional turmoil and Jim was trying to make him feel better. Jim felt compassion for George. He offered to do some repair work on the script and he needed a copy of George's original script. George asked me to accompany him on his trip back to Jim's home. I did just that, and George gave Jim a draft of his original script.

When we returned to George's home, George said, "I was very happy with the work Tom was doing on my script. He was doing a good job. I wish I had some way to get in touch with him. Bob didn't want us to exchange phone numbers, so I can't call him. He wanted us to communicate only through him."

"I think Tom gave me his card when he was here for your mother's party," I told George. "I may still have it." I went upstairs and found Tom's card. George immediately called Tom. Tom was surprised to hear from George. That evening, Tom came to George's home and brought research work that he had done for George. Tom told us that he never wanted to see Bob Munger again. He also told George that he believed in everything George was doing and he wanted to help George complete the script. Their rewrite is now complete, and I have read it.

It was almost a year later when Tom told me the reason he no longer liked Munger. "Bob was taking money from George with no concern for George's story," Tom said. "Bob told me not to listen to anything George said. At each meeting, George told me the points in his story that had to remain intact. As soon as the meetings ended, Bob told me to not listen to George. Bob told me to throw out most of the things George wanted to remain in the script. Bob said George was a rich man who wanted a movie script. But, at one meeting I heard George tell Bob he had borrowed the money for the rewrite, so I knew Bob was lying. I also saw the devotion George had for his mission and his story. It was his life. Bob was ripping George off and I couldn't be a part of it. I had to get away from Bob and I had no way to contact George."

I have met Tom several times since the script has been rewritten. Tom has great respect for George and strongly believes in George's mission. He also refused to take any money for helping George. "My reward is knowing I helped George fulfill his mission," Tom said.

Many times I have heard George say, "Thank God I was able to contact Tom, he saved my story."

I affirm George's credibility and I testify that all the statements on this page are true.

Sincerely,

John B Buonomo

John Buonomo

279

March 5, 2004

Sharon Younker
~~■~~
P.O. Box ■
Taylorville, Illinois 62568

Documentation for George Newberry's Story

To Whom It May Concern:

George Newberry and I have been friends for over fifty years. I went to Junior High and High School with George. We graduated together in the Taylorville High School Class of 1958.

George and I have always had a very close friendship. He now lives in California, but we talk on the phone three for four times a month. George has always cared about his hometown friends.

In 1976, I received a call from Marilyn Malmberg. She was flabbergasted. She told me she had just received a call from our mutual friend, George.

"Sharon, I can't believe it," she said. "George told me when he woke up he saw LaDonna floating over his bed. He had no way of knowing she died."

Marilyn and I were both amazed by that incident. Neither Marilyn nor I had talked to George for quite some time. George could not have known LaDonna died the same morning he saw her floating above his bed.

When we were in Junior High School, George was very ill. I remember him telling everyone that during his illness, he had a near-death experience. From that experience, he has always said that he has a destiny to fulfill. I believe he does.

George has an excellent sense of humor. He loves to laugh and joke, but he is very spiritual. I have known George for many years and I know him well. He does not lie. Over the years, George has told me everything that has happened to him. I know George's story is hard to believe, but I also know it is true.

I testify that every statement I have made in this documentation is the truth.

Sharon Younker
Sharon Younker

State of _Illinois_ County of _Christian_
On _3-8-04_ Before me _Cheryl L. Cockrell_ Notary Public, personally appeared _Sharon Younker_

she proved to me on the basis of satisfactory evidence to be the person whose name is subscribed to the within instrument and acknowledged to me that she executed the same in her authorized capacity, and that by her signature on the instrument the person or the entity upon behalf of which the person acted executed the instrument.

Witnessed my hand and official seal

Cheryl L. Cockrell
Notary Public

"OFFICIAL SEAL"
CHERYL L. COCKRELL
NOTARY PUBLIC STATE OF ILLINOIS
MY COMMISSION EXPIRES 6/14/2004

280

March 5, 2004

Marilyn Malmberg
▬ N. ▬▬ St. # ▬
Taylorville, Illinois 62568
(217) ▬▬-▬▬

Validation For George Newberry's Story

To Whom It May Concern:

I have known George Newberry for almost 60 years. George and I have been friends since we were four years old. We went to the same Church and we graduated together from Taylorville High School with the Class of 1958. George moved to California in 1968, but we have maintained a very close friendship and we always keep in touch.

In October of 1976 I received an amazing phone call from George. He called my parent's home to ask them for my phone number. To our surprise, I happened to be there and I answered the phone. At that point in time, I hadn't talked to George for at least two years. George immediately told me why he was calling.

"Marilyn, I'm calling you because something very strange happened. When I woke up this morning, I opened my eyes and I couldn't believe what I saw. LaDonna Higdon (One of our classmates) was floating above my bed. She didn't say anything. She was smiling at me. Then, within a few seconds, she disappeared. I know I saw her, it wasn't my imagination. I was awake."

I was shocked. I couldn't believe what George had said. "Oh my gosh, George," I replied. I just found out LaDonna died."

We were both amazed by what had happened. Over two thousand miles away, George had seen LaDonna's spirit floating above his bed, on the same day she passed away. LaDonna died at the age of 35, on October 14, 1976.

As soon as my conversation with George ended, I called my best friend, Sharon Younker. I couldn't believe what had occurred and I had to tell someone. Sharon and I both know George couldn't have known LaDonna died before I told him.

In 1994, I spent two weeks at George's California home. He had just finished the first draft of his script. The title is now The Devil's Reign, but at that time it was False Prophets

While I was in California, I went with George to several meetings with people who were trying to get the film produced. I met Shirley Krims and Antoinette Meyer. George mentions both of them in his book. Shirley believed in George and his story, and was trying to find someone to help him. She introduced George to Antoinette. I was present when Antoinette told George she had the connections to get the movie made and it was a done deal. Unfortunately, her statements proved to be false and nothing happened.

I can also verify that in the 9th Grade, George was very ill and missed months of school. That was in 1955. At that time, George told his friends that after his surgery he had a near-death experience. Since then, George has talked about that experience many times.

George is a good person. He is honest, he cares about other people and he is a good friend. I know his story is true.

I testify that all the statements I have made in this documentation are true.

Marilyn Malmberg
Marilyn Malmberg

State of _Illinois_ County of _Christian_
On _3-8-04_ Before me _Cheryl L. Cockrell_ Notary Public,
personally appeared _Marilyn Malmberg_.

she proved to me on the basis of satisfactory evidence to be the person whose name is subscribed to the within instrument and acknowledged to me that she executed the same in her authorized capacity, and that by her signature on the instrument the person or the entity upon behalf of which the person acted executed the instrument.

Witnessed my hand and official seal.

Cheryl L. Cockrell
Notary Public

"OFFICIAL SEAL"
CHERYL L. COCKRELL
NOTARY PUBLIC STATE OF ILLINOIS
MY COMMISSION EXPIRES 5/14/2004

281

March 2, 2004

Edie Boudreau
P.O. Box ▓
Eureka, Nevada 89316-0594
(775) ▓▓▓-▓▓▓▓
(949) ▓▓▓-▓▓▓▓

RE: George Newberry Story Validation

To Whom It May Concern:

I have known George Newberry since 1983. I met him at a writer's club meeting. He was searching for someone to help him write his story. His story was quite interesting and I volunteered to help him

When I met George, only a few unusual things had happened to him. He told me about his near-death experience, the dream that started his mission, and his awakening to see the spirit of LaDonna floating above his bed, on the day she died. I didn't realize I would later become personally involved in one of George's strange experiences.

We worked on George's story for only a short time before circumstances put it on hold. However, we had developed a lasting friendship.

In mid 1996, I went to George's home to visit. He told me his friend, Shirley Krims, had wanted to watch the videotape of a television interview he had done in North Carolina. When George inserted his tape into her VCR, it ate up his tape and her VCR broke. George had retrieved the damaged tape and he had taken Shirley's VCR to a repair shop.

I told George, "Give me your tape and I will make a copy for you. You should always show people your copy. If anything should ever happen to it, you will still have your master tape"

A few days later, I returned the original tape and a new copy to George. That evening I received a call from George. He was in Los Angeles. "Are you trying to pull a joke on me?" he asked. "What do you mean?" I replied. George exclaimed, "I played the tape for Shirley, and there are strange sounds on the soundtrack that weren't there before, and there's an evil voice saying, "Lies, Lies, Lies."

I didn't know what George was talking about. I had to hear it for myself. I met George at his home in Ontario at 11:30 that evening. I told him I had made two copies of the tape, one copy for him and another for me, so I could watch the show later. While I waited for George to return home from Los Angeles, I watched my copy. There was nothing unusual in the soundtrack. When I made the copies, I had no time to watch the tape. I didn't turn my TV on. I placed the original tape in one VCR and made the copy on another. At random, I picked up one copy of the tape and gave it to George. When I watched the tape I had given to George, I was astonished. There were unusual sounds on the tape, and an evil voice repeated, "Lies, Lies, Lies." About 30 seconds later, the same voice mumbled, "My daughter learned how to speak Spanish." That's when I realized what had happened. "I have a neighbor who is a drug dealing scumbag," I said to George. "He has a CB Radio. I don't know how it happened, but my VCR picked up his signal. I'm almost positive that's his voice."

I later confirmed the voice was his. He was eventually arrested on a drug charge and he spent time in prison. He still has no knowledge that his voice was recorded on the tape. It is strange that, "Lies, Lies, Lies," was recorded at the most pertinent spot on the tape.

I affirm that George is a man of good character and his story is true. I am willing to take a Polygraph test to validate my statements.

Sincerely,

Edie Boudreau

Edie Boudreau

State of _CALIFORNIA_ County of _SAN BERNARDINO_
On _March 2, 2004_ Before me _____ Notary Public,
personally appeared _EDIE M. BOUDREAU_

She proved to me on the basis of satisfactory evidence to be the person whose name is subscribed to the within instrument and acknowledged to me that she executed the same in her authorized capacity, and that by her signature on the instrument the person or the entity upon behalf of which the person acted executed the instrument.

Witnessed my hand and official seal. _____

Notary Public

March 24, 2005

Mr. Jim Hardiman
▉▉▉ ▉▉▉▉▉▉▉
Cathedral City, CA 92234
(760) ▉▉▉-▉▉▉▉

RE: Validation for George Newberry's Story

To Whom It May Concern:

In August of 1997, Robert Munger brought me a partially completed script titled "False Prophets." He asked me to rewrite and complete it for him. He told me to maintain a plot about a girl who was being pursued by the forces of evil. He gave me carte blanche to write any story I wanted.

In October of 1997, I completed the script. Bob came to my home in Cathedral City, California. George Newberry accompanied him. That was the first time I knew George existed. Bob had never mentioned anything about George to me. Bob handed George the script I had written and told him to read it while Bob and I visited.

When George finished reading the script, Bob asked him what he thought about it.

George replied, "If someone wants to make a good horror film, it's a good story. But, it has nothing to do with my script or my treatment. This is supposed to be a polish of my script, not a totally new story."

"What script and what treatment are you talking about?" I asked.

George told me about his near-death experience, his mission, and the supernatural experiences that he encountered while working on his script. The script was supposed to tell George's story and relay his message to his audience. Bob did not tell me to polish or rewrite an existing script. I didn't know George's script or treatment existed. George had every reason to be upset. His story was gone.

I asked Bob why he hadn't told me about George's script or treatment. "I thought I gave his script and treatment to you," Bob replied.

Bob had never mentioned George Newberry, his script or his treatment, to me. As a result, all of the effort I had put into writing that script was wasted.

I agreed that George's story had to remain intact. Otherwise, the messages people were supposed to get from seeing his movie were lost. I understood why George was upset. I wanted to help him save his story so I offered to revise the script. Bob said he would send me copies of George's material.

The next day I received many calls from George. He was very upset because Bob had caused his story to be destroyed. I understood and I tried to console him. Instead of waiting for Bob to send George's material, I told George he could bring it to me personally. George desperately wanted to save his story and I wanted to help him. George brought me copies of his script and treatment and I did my best to revise the script I had written.

George and I still share a friendship. The purpose of this letter is to validate that everything George Newberry wrote in his book that concerns my interaction with George, Bob Munger and George's script, is the truth.

Sincerely,

Jim Hardiman

State of CALIFORNIA   County of San Bernardino
On March 25, 2005 Before me Irene Rose Guerrero Notary Public, personally appeared Jim Hardiman .

He proved to me on the basis of satisfactory evidence to be the person whose name is subscribed to the within instrument and acknowledged to me that he executed the same in his authorized capacity, and that by his signature on the instrument the person or the entity upon behalf of which the person acted executed the instrument.

Witnessed my hand and official seal.

Irene Rose Guerrero
Notary Public   Notary Public

Copy of the letter sent to me by David Hiatt.

David Hiatt

Literary Agency

██

February 27, 2003

████████████

Dear George:

I have enclosed a full corrected manuscript of *The Devil's Reign*. After I receive your letter indicating your satisfaction with the editing work we can begin the marketing effort.

I think there is something dark and unusual with this project. I have had numerous computer problems which appeared only when working with your story. The last one completely wiped out the software and the computer had to be rebuilt from scratch. I hope the devils are behind us.

Best regards,

David Hiatt

Printed in the United States
96062LV00009B/147/A

9 781420 853988